Hydraulic Pumps and Motors

FLUID POWER AND CONTROL

A Series of Textbooks and Reference Books

1. Hydraulic Pumps and Motors: Selection and Application for Hydraulic Power Control Systems, *by Raymond P. Lambeck*

Other Volumes in Preparation

Designing Pneumatic Control Circuits: Efficient Techniques for Practical Application, *by Bruce E. McCord*

Hydraulic Valves and Controls: Selection and Application, *by John J. Pippenger*

Fluid Power Troubleshooting, *by Anton H. Hehn*

Hydraulic Pumps and Motors: Selection and Application for Hydraulic Power Control Systems

Raymond P. Lambeck

Sperry Vickers
Troy, Michigan

MARCEL DEKKER, INC. New York and Basel

Library of Congress Cataloging in Publication Data

Lambeck, Raymond P.,
Hydraulic pumps and motors.

(Fluid power and control ; v. 1)
Includes index.
1. Fluid power technology. 2. Hydraulic control.
3. Pumping machinery. 4. Hydraulic motors. I. Title.
II. Series.
TJ843.L25 1983 621.2'52 83-5342
ISBN 0-8247-7014-5

MARCEL DEKKER, INC.
270 Madison Avenue, New York, New York 10016

Current printing (last digit):
10 9 8 7 6 5 4 3 2 1

PRINTED IN THE UNITED STATES OF AMERICA

Series Introduction

During the past decades, four major driving forces, acting concurrently, have combined to transform fluid power from a leaking combination of relatively primitive controls and actuators to a science which combines sophisticated fluid power actuators, intelligent controls and energy efficient power sources with complex electronic devices. The result is an electronically controlled fluid power system capable of powering almost every imaginable machine.

The first of these forces has been the emergence of applied research aimed at defining, measuring and improving the performance of the various components used in fluid power systems.

The second force has been the impact of suddenly increased energy costs. This resulted in the development of energy efficient devices, made the use of variable displacement pumps more economically attractive, spurred the adoption of new systems of fastening and sealing fluid connectors and conductors and forced development of components which can survive when working with high water content fluids.

The third force has been the rapid development of variable speed electrical motor drives with small motors up to 3 to 5 HP units coupled to a ball screw mechanism or harmonic drive device applied to machine motions which in the past had been solely in the fluid power domain.

The fourth major force has been the rapid development of reliable, factory compatible electronic controls. It is virtually mandatory that the fluid power industry keep pace with the electronic revolution by development and producing real products which will continue to reflect the current state of the art for fluid power technology.

The essential purpose of the Fluid Power and Control series is to provide an overview of fluid power products, including their construction and application, to reflect the current state of the art of the technology. The material presented should be useful to both component designers in accelerating their efforts in dynamic product design as well as to engineers as they work to optimize the performance of all types of machines through the application of the newest fluid power products and science.

Contributors to this series have been encouraged to discuss component designs and functions which are currently in use as well as the leading edge components which are just becoming available with only brief reference to yesterday's designs and applications. Frequent reference to load sensing, variable displacement pumps and other energy saving designs, electrically positioned controls and actuators and means of minimizing fluid and energy losses provide the reader with insight into various methods of improving the performance of his components and systems.

This volume, Hydraulic Pumps and Motors, has been written by an acknowledged expert with a vast amount of experience and data made available to him from Sperry Vickers, a major producer of almost all types of hydraulic pumps and motors for many years. This exposure has enabled the author to provide a definitive work on the subject and makes this volume an especially worthwhile addition to any engineering library where information on power transmission is to be found.

Z. J. Lansky

Preface

In recent years, hydraulics has been called upon to provide intelligent muscles for the machines and vehicles of the modern world. Its many advantages of compactness, flexibility, responsiveness, and controllability have been applied by the machine designer in innovative ways to power a multitude of functions smoothly, precisely, and efficiently. Today, to achieve his ends, the system designer has available to him a wide variety of pumps, controls, and other system components developed to provide optimum performance under very exacting demands. The proper choice of hydraulic elements will provide a machine with the responsiveness, productivity, and reliability demanded by the ultimate user.

It is the purpose of this book to review the primary types of pumps, motors, and controls in widespread use today. While the internal construction of the various types will be discussed, it is not for the purpose of going deeply into all of the subtleties and intricacies of component design. Rather, it is to convey some appreciation for the advantages (or limitations) of certain component designs in various types of operation.

Anyone who has been involved in design work of any kind recognizes that there is no such thing as the "perfect" design. The demands of size, weight, strength, durability, efficiency, producibility, and installation flexibility are often in conflict, and the optimum combination of internal design features is often hard for the component designer to find. However, fortunately for the user of hydraulic power, there are many highly refined products available today from which to choose, each with its own preferred field of application. The information in this book is aimed at helping the system designer to select and apply hydraulic pumping units in an efficient and effective manner to produce responsive and reliable systems.

The world of hydraulic applications, aside from aircraft and a few other specialized cases, can be divided loosely into two groups. One is industrial hydraulics, where most machines are located in a protected environment and hydraulic power units are driven by synchronous electric motors. The other is mobile hydraulics, where all-weather operation is usually required and hydraulic power units are driven by internal combustion engines.

While the demands of these two fields of application can be quite different, most of the pump and motor types described in this book are sufficiently versatile to span both areas. In some cases, the designs may be individually modified to satisfy more ideally the needs of the ultimate user in his particular field of application.

The types of pump and motor controls and the types of systems described are also found in both fields.

It is hoped that this book will prove of value to any engineer involved in hydraulic system design. It will remind the more experienced machine or vehicle designer of the broad range of products, special controls, and system configurations available. It will also reemphasize some of the other installation factors that must not be forgotten in assuring reliable service. The twin concerns of energy conservation and maximum machine productivity demand a thorough job of component selection and system planning. All options must be understood and evaluated.

Even more than to the experienced engineer, though, this book is directed toward the newcomer to the hydraulics field and to the engineer who has only occasional need to be concerned with hydraulic systems. The information presented is straightforward and practical, requiring little in the way of prior knowledge. The text is general and non-mathematical. The intent is to provide a practical foundation for the systems engineer and a ready-reference for future use.

Hydraulics is a diverse and interesting field. It is growing daily and there is still room for innovation. It is our purpose to lay the foundation for the next surge forward, the next generation of machines. The ultimate system has not yet been designed. Engineers, here is your opportunity!

Raymond P. Lambeck

Acknowledgments

The preparation of this book required access to information from many sources and the assistance of many people.

Thanks is particularly due for information received on high-torque low-speed motors from Bignozzi in Italy, National Hydraulics in Canada, Riva Calzoni in Italy, and Renold Ltd. in England.

Other valuable and useful data were contributed by the Sperry Vickers operations in Bad Homburg and Hof, Germany, in Tulsa, Oklahoma, and by the Sperry Vickers Japanese affiliate, Tokyo Keiki Co.

Most of all, thanks is due to Sperry Vickers in Troy, Michigan, for unlimited access to laboratory data, design information, and product illustrations.

While it is impossible to list all of the individuals who contributed data and assistance, there are three in particular whose key contributions deserve special recognition. The first is Doug Drummond, who encouraged me to start the book and helped put it in final form. Second, the book itself could not have been completed without the care and interest of Loch Macadam in preparing the many drawings, graphs, and diagrams. Finally, the organization and preparation of the final manuscript was possible only through the special efforts of Karen Nicholas who kept everything in order and saw that the work was complete and accurate.

Contents

Hydraulic Pumps and Motors

1
History

I. INTRODUCTION

The evolution of modern hydraulic power systems required the development and refinement of four interdependent elements: 1) pumps to generate the hydraulic power; 2) motors and actuators to convert that power to the desired use; 3) transmission lines and valves to distribute and control its application; and 4) the development of a body of engineering theory to guide efficient and productive use of the above in complete power systems.

Progress in each of these four parallel directions has been going on fitfully and unevenly for many centuries, often in totally unrelated fields. It has only been in relatively modern times that these four elements have been brought together to produce systems capable of operating efficiently at high power levels with rapid response and absolute precision of control.

II. EARLY ORIGINS

The history of hydraulic power is a long one, dating from man's prehistoric efforts to harness the energy in the world around him. The only sources readily available were the water and the wind—two free and moving streams.

The watermill, the first hydraulic motor, was an early invention. One is pictured on a mosaic at the Great Palace in Byzantium, dating from the early fifth century. The mill had been built by the Romans. But the first record of a watermill goes back even further, to around 100 BC, and the origins may indeed have been much earlier. The domestication of grain began some 5000 years before and some enterprising farmer is bound to have become tired of pounding or grinding the grain by hand. Perhaps, in fact, the inventor was some farmer's wife, since she often drew the heavy jobs.

Many mills stayed in use until the close of the 19th century, but they had been in the process of being replaced by the steam engine as a source of power for the previous 150 years. The transmission of the power generated by the waterwheels was by shafting and crude gears or pulleys. The steam engine for the first time employed an enclosed moving stream of fluid (steam) under pressure—the principle to be later employed in hydraulic power transmission—for transmitting power continuously from the point of generation to the place where it was used.

The transmission of hydrostatic power to a distant point began long before that, however. Hero of Alexandria, in the first century AD, built a device in which a fire on the temple altar expanded air in a closed container. The air pressure forced water to travel along a tube to the temple doors where, spilling into a container, it provided the force through ropes and pulleys to magically open the doors.

Hero also produced a steam engine of sorts. Its motion depended on the reaction forces from jets of steam, as in the present steam turbine, but it was little understood and it was no more than an interesting curiosity.

The invention of early forms of the hydraulic pump has a similarly early origin. The first pumps were not used to develop hydraulic power, however, but only to transfer water for irrigation or to remove it from mines.

Archimedes applied the principle of the screw to hydraulic machines in the third century BC. His screw pumps were used to raise water for irrigation or to the level of aqueducts. The piston pump, the first mechanical device capable of generating pressure in a column of liquid, is believed to have come from Egypt at a similarly early date. Like the screw pump, it was used only as a means for moving water and not as a means for generating hydraulic power.

III. DEVELOPMENT OF HYDRAULIC THEORY

The theory needed to guide the application of hydraulic power was slow in coming. When Archimedes discovered that an object's weight in water is reduced by the water it displaces, he actually discovered the effect of hydrostatic pressure. The theory that explained this phenomenon was not developed until more than 1800 years had passed. It was not until around 1600 that Stevinus, a Dutchman, showed that hydrostatic pressure varies with the depth of the fluid and is independent of the shape of the vessel.

The 17th and 18th centuries were a productive period in the development of hydraulic theory. They laid the groundwork for the development of hydraulic power systems which was to follow.

Torricelli studied fluid motion in the early 17th century. Late in that century, Sir Isaac Newton conducted studies on viscosity and the resistance of submerged bodies in a moving fluid.

The key achievements of the period occurred in the middle of the 18th century when Daniel Bernoulli developed the theory of transmission of energy in fluid streams and Blaise Pascal, at about the same time, established the principle that pressure is transmitted equally and undiminished in all directions in an enclosed fluid. The use of this principle to create the force multiplying effect of pressure on large areas was first used in large presses in the latter part of the 18th century.

The fundamentals of fluid theory were established by the above work and refinements were added by Navier who derived the mathematics of motion in liquids including equations for fluid flow with friction. This was early in the 19th century. It was followed by the work of Stokes, who independently discovered the same equations and further extended the work of Navier.

An important added contribution was made by Osborne Reynolds, late in the 19th century, through his fundamental studies of resistance to flow and the phenomena of laminar and turbulent flow.

All of the above theory was based on the mathematical forms developed by Descartes, Newton, Euler and Laplace.

A final summing up of all of this work, plus some important individual contributions was made by H. Lamb, a student of Stokes. His definitive work, "Hydrodynamics," was first published in 1879 as "A Treatise on the Motion of Fluids." It was later enlarged in 1895, and in subsequent years through six editions. It is a comprehensive work on the subject, collecting the work of Lamb himself, and of other contributors of his time.

IV. LUBRICATION

Lubrication, too, has its history and its theory. Chariot wheels were lubricated with beef or mutton tallow before 1400 BC. Even petroleum refining has ancient origins. It was described by Herodotus in the 5th century BC. There are also references to vegetable oils in very early writings. A true understanding of the way lubricants act did not develop until much later, however.

Modern hydrodynamic lubrication is based on the law, discovered by Newton, that the shearing stress is proportional to the rate of shear. This relationship, between shear rate and shear stress, is called viscosity. Early experimenters such as Amontons (late 17th century), Coulomb (late 18th century), followed by Him, Thurston and Petroff in the 19th century developed empirical laws for friction without fully understanding or applying the law of Newton.

The beginning of our modern understanding of lubrication occurred when in 1883, Beauchamp Tower discovered the buildup of pressure in the converging film of a journal bearing. Osborne Reynolds interpreted this as a hydrodynamic action and developed the equations governing the generation of pressure in the film.

Soon after, Kingsbury applied the principle of a converging film to a thrust bearing by providing a series of tilting segments.

While various types of antifriction bearings with rolling contact are widely used today, the hydrodynamically lubricated bearing surface still plays a key part in pump and motor design as well as in other machinery. In pumps and motors, it is possible to draw on the fluid medium to lubricate the moving surfaces. In well-conceived designs, the surfaces are separated by a fluid film and there is no physical contact during normal operation, indefinitely extending the life of the unit.

V. POWER TRANSMISSION

The development of a practical steam engine by Thomas Newcomen in the early 18th century created the need for transmitting sizable amounts of power, sometimes to points at some distance from the engine. There was some use of underground water lines, carrying water under moderate pressures, as a power distribution system. The engineers who developed underground water power probably felt that they had invented and established the ultimate in long range power transmission—and indeed they had for the time. But the advent of electrical power generation and distribution soon made the earlier system obsolete.

Hydraulics today is usually concerned with the transmission and distribution of power over relatively short distances, commonly measured in feet. The fluid is rarely water. The introduction of oil hydraulic fluids with good lubricating properties coupled with advances in metallurgy have made higher operating pressures and the generation and transmission of hundreds of horsepower practical.

An early system of surprising sophistication was the hydrostatic transmission installed to elevate the guns of the battleship "Virginia" in 1906. It combined both the high power capability and the fine precision of hydraulics in a system that was far ahead of its time. Many other hydraulic applications for naval and industrial use followed close behind.

VI. DYNAMIC CONTROL

The development of small, light and efficient power units was accelerated by the use of hydraulic power in aircraft applications, beginning with such auxiliary functions as raising the landing gear. The use of hydraulic power for purposes of control saw its first widespread use in the aeronautical field in the constant speed, variable-pitch propeller. The first commercial application was on the Boeing 247D twin engine transport in the early 1930s. It appeared thereafter in the Douglas DC-3 and all subsequent Douglas, Boeing and Lockheed transports.

In the constant speed propeller, the force needed to move the propeller blades against the centrifugal and aerodynamic moments was supplied hydraulically. The pitch was automatically controlled during takeoff, climb and the remainder of the flight to regulate engine speed while providing maximum propeller thrust. Hydraulic power assist for aircraft controls first appeared in the Douglas DC-4 airplane in the late 1930s. From that point on, as aircraft grew larger and flew faster, powered flight controls became an essential part of all new designs. The Boeing 747, for example, has 500 hydraulic horsepower available. The demands for light weight, reliability, efficiency and precision of control did much to accelerate component development and spur the advance of modern hydraulics.

VII. HYDRAULIC SYSTEM DEVELOPMENT

In more recent years, the role of leadership in hydraulic power application has been taken over largely by some of the larger earthmoving and construction equipment manufacturers. The total power involved is often greater than that required in even the largest aircraft systems. The concentration of this power in a few very large loads, as compared to the multitude of smaller loads in an airplane, has spurred the system designer to find new, creative ways to distribute and control the power. This he has succeeded in doing to obtain maximum precision of control and productivity while minimizing power consumption.

In modern mobile and industrial applications, the system designer has made full use of the wonderful flexibility and adaptability of hydraulics. It can readily be directed around corners and past moving joints that virtually defy mechanical power transmission. It can be divided easily to serve individual loads or recombined to power a high, single-load demand. Electrical and pneumatic power

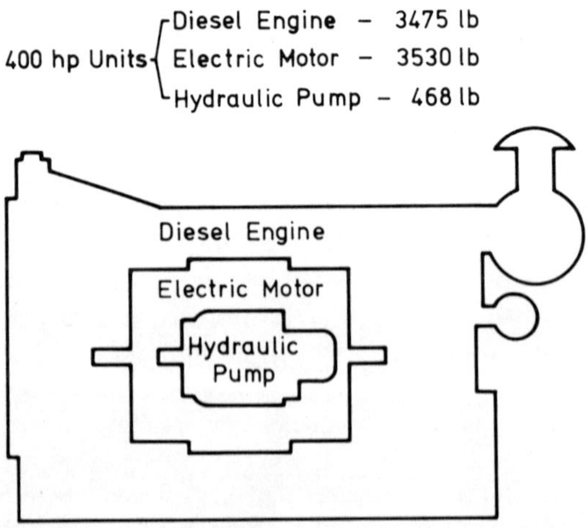

FIGURE 1-1 Comparative Size of Diesel, Electric and Hydraulic Power Units

distribution have similar flexibility, but lack the compactness and certain of the other important characteristics of hydraulics.

A very significant feature of hydraulic power is its extreme power density. Figure 1-1 shows a comparison of a 400 hp hydraulic pumping source with a diesel engine and an electrical motor of the same power capacity. The contrast is obvious and is one that offers both an opportunity to the user of hydraulic power and a challenge to the designer of hydraulic components.

The high-power density in a hydraulic pump or motor creates design challenges beyond those found in many other products. High loads on rapidly sliding surfaces, extreme fluid velocities and rapidly recurring pressure applications on pumping elements (vanes, pistons, gears) demand careful design attention to assure long life, complete reliability and efficient operation. The user, on the other hand, reaps the benefit of a family of apparatus that is highly responsive, finely controllable and simple to apply. When the designer understands the user's needs and the user understands the equipment's full capabilities, the maximum potential of hydraulic power can be achieved.

2

Fixed Displacement Hydraulic Pumps

I. INTRODUCTION

Over the years, a remarkable variety of pumping concepts has been invented and produced. Some have successfully survived the changing and increasingly stringent demands that have come with the growth of hydraulic power. Others have been adapted to specialized needs. Many, however, have been found unequal to modern needs and have either disappeared from the market or found little application. In true Darwinian fashion, only the fittest have survived. The key to a successful design is not always obvious, but the following factors are usually involved.

In the surviving designs, the areas subject to pressure are held to a minimum or are hydrostatically balanced to minimize contact forces and bearing loads. Opportunities for internal leakage across sealing surfaces are minimized. Cooling flow across sliding or moving surfaces is adequate to prevent local overheating. Thermal distortion and operating deflections are allowed for while still maintaining running clearances and bearing alignment; and finally, the designs are relatively simple and can be produced economically.

The pumps described in the following section are the types that have survived the test of time and have captured an overwhelming share of the hydraulic power applications. Each type is made by numerous manufacturers, each having exercised his own choice of internal design details. Among them, they cover virtually every conceivable hydraulic need.

Since the subject of the book is power hydraulics, the designs described are those having a capability of producing at least 1000 psi (70 bar), although they may sometimes be used at lower pressure. Pumps with more limited pressure capability sometimes find auxiliary use in power circuits for such purposes as transmission supercharge pumps, but are rarely used alone above fractional power levels.

II. COMPARISON OF INDUSTRIAL AND MOBILE HYDRAULIC POWER DEMANDS

The characteristics that distinguish industrial hydraulic system demands from mobile vehicle demands are principally:

Industrial	*Mobile Vehicle*
Moderate drive speeds (1200-1800 rpm range with constant operating rpm in any given application)	High drive speeds (2000-3000 rpm with variable operating rpm down to 500-600 rpm)
Pressures below 1000 psi for machine tools and most frequently below 3000 psi for other applications.	Pressures generally above 2000 psi and many applications in the 5000 psi range.
Moderate fluid temperatures (< 160°F).	High normal operating temperature (160°-200°F) with excursions to 225°F.
Moderate ambient temperature.	Extreme range of ambient temperatures.
Environments are, for the most part, relatively clean.	The surroundings often include high concentrations of dust and dirt.
Strong emphasis on low sound level: 70 dB(A) desirable; 80 dB(A) mandatory	Less emphasis on sound (must be below diesel sound level): up to 90 dB(A) may be acceptable.
Non-petroleum fluids include 95:5 soluble oil and water, invert emulsions and occasionally water/glycol or synthetic fluids.	Non-petroleum fluids include synthetics and water/glycol.

III. BALANCED VANE PUMPS

A. Operating Principle

The balanced vane pump is a very versatile design which has found widespread use in both industrial and mobile applications. The basic design principle is shown in Figure 2-1. The rotor and vanes are contained within a double eccentric cam ring and there are two inlet segments and two outlet segments during each revolution. This double-pumping action not only gives a compact design, but also leads to another important advantage: although the pressure forces acting on the rotor in the outlet area are high, the forces at the two outlet areas are equal and opposite, completely cancelling each other. As a result, there are no net loads on the shaft bearings. Consequently, the life of this type of pump in many applications has been exceptionally good. Operating times of 24,000 hours or more in industrial applications are widespread. In the more severe conditions encountered in mobile vehicles, 5,000 to 10,000 hours of trouble-free operation is frequently achieved.

1. Intravane Feature Many of the balanced vane pumps also have an intravane feature which helps to keep the vane in continuous contact with the cam ring during the rises and falls while also minimizing vane-tip contact forces. In combination with the vane-tip design, the intravane also serves to balance the radial pressure forces on the vane during the pumping action. As seen in Figure 2-2, when the vane is at full extension, the pressure force on the leading edge of

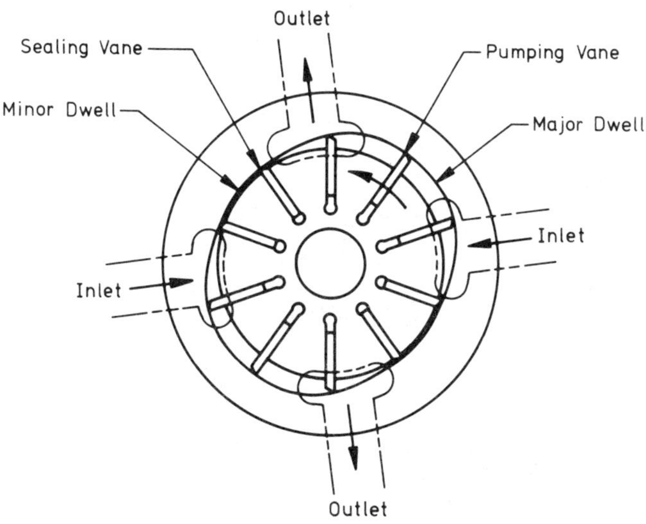

FIGURE 2-1 Balanced Vane Pump Operating Principle

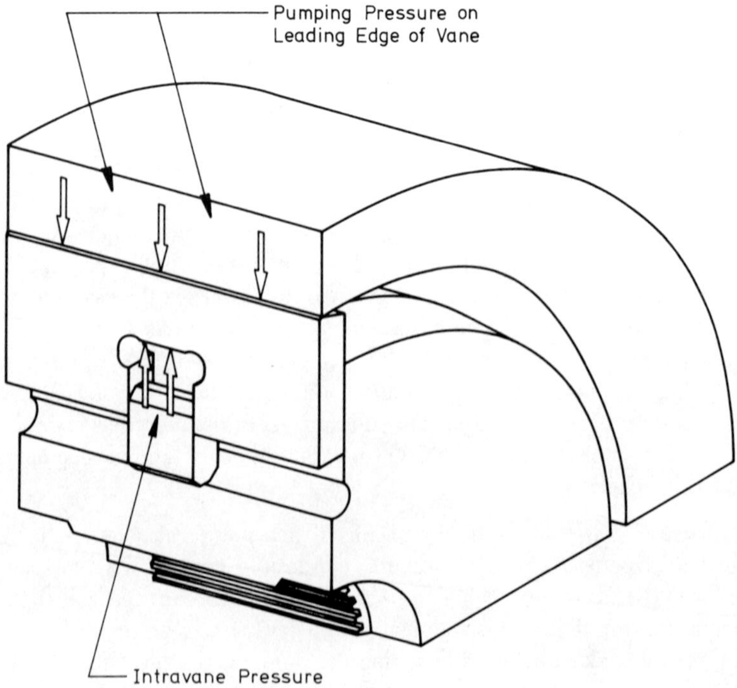

FIGURE 2-2 Pressure Balance on Pumping Vane

the vane is countered (and slightly overbalanced) by the pressure force in the intravane area. When the vane is at minimum extension, as shown in Figure 2-3, the pressure on the trailing edge of the vane is countered (and again slightly overbalanced) by the pressure in the full undervane area. By careful design of the vane-tip shape and the intravane size, the net forces at the vane tip can be kept small at all times and wear during normal operation is extremely low.

2. Constant Delivery Feature Figure 2-1 illustrates one other feature of the balanced vane design. Pumping occurs while one vane is in contact with the major dwell and one with the minor dwell of the cam ring. Each dwell is essentially a constant radius so the fluid delivered per degree of rotation is a constant value. As a result, the flow delivered by a balanced vane pump is very smooth compared to most other types of pumping mechanisms, and sound levels are inherently low.

B. Balanced Vane Pump Sound Levels

Careful attention to the manner in which pressure is picked up by each successive vane also enhances the smoothness and quietness of operation. Figure 2-4 shows the low sound levels achieved in modern industrial vane pumps.

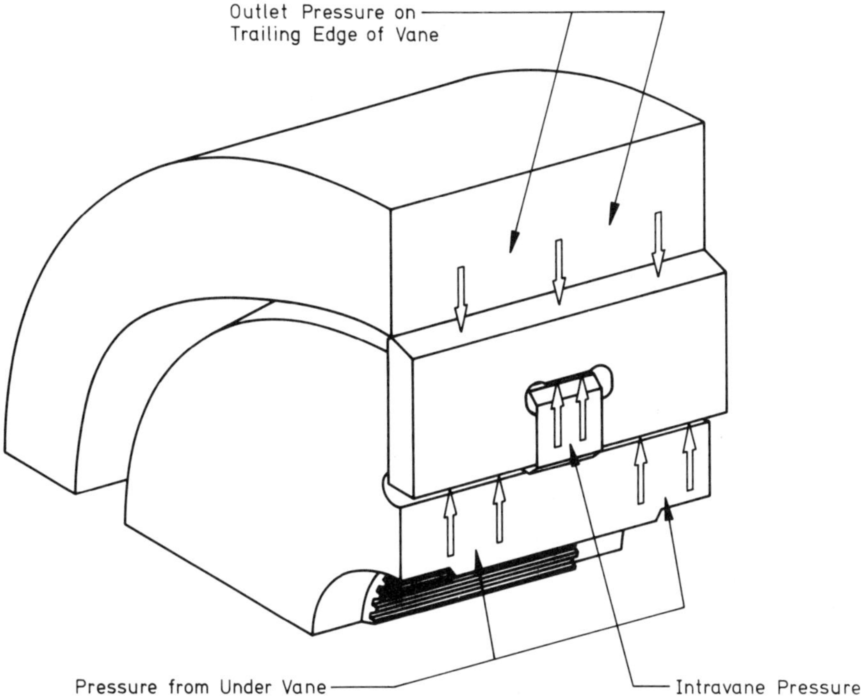

FIGURE 2-3 Pressure Balance on Sealing Vane

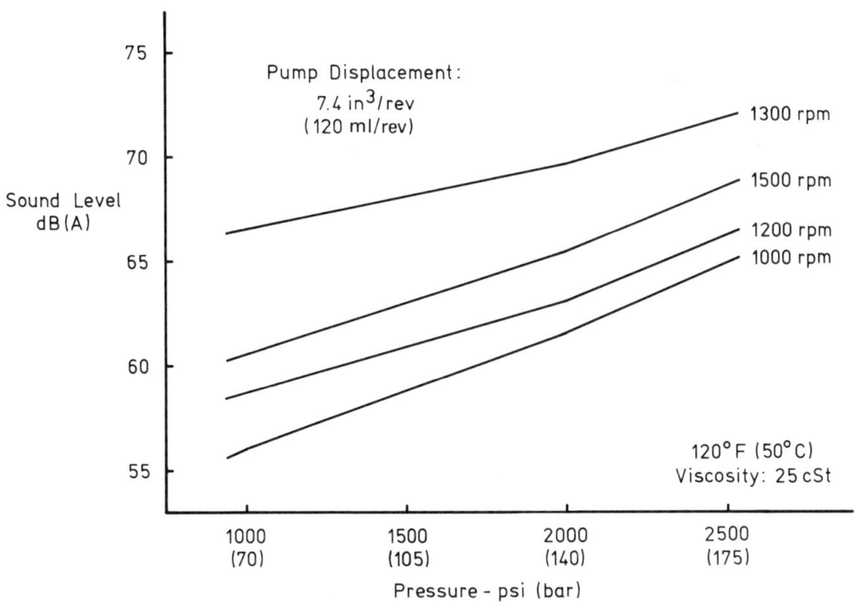

FIGURE 2-4 Balanced Vane Pump Sound Levels

Most industrial vane pumps in the United States are run at 1200 rpm and at pressures between 1000 and 2000 psi (70 to 140 bar). Over this range, the sound level of the pump shown is 63 dB(A) or less as measured using NFPA Standard procedures. (The sound level of normal speech is 60 dB(A).) The pump delivers 40 hp (30 kW) of hydraulic power at 1200 rpm and 2000 psi (140 bar). Even at 1800 rpm and 2000 psi (140 bar), where the pump is delivering 60 hp (45 kW) of hydraulic power, the sound level is only 70 dB(A).

C. Balanced Vane Pump Performance

Typical performance of balanced vane pumps is illustrated in Figures 2-5 and 2-6. Ratings generally range up to 1800 rpm and 3000 psi (210 bar) at 150°F (65°C) for industrial units. For mobile-vehicle units, speeds run in the 2000 to 2500 rpm range, depending on size, with pressures to 4000 psi (280 bar) and temperatures to 200°F (95°C) continuous and 225°F (110°C) for brief excursions.

1. Pressure-Loaded Flex Plate Construction To maintain high volumetric efficiency and also withstand the dimensional changes that occur due to thermal excursions, one type of balanced intravane pump is equipped with flexible pressure-loaded plates which automatically adjust to the expansion and con-

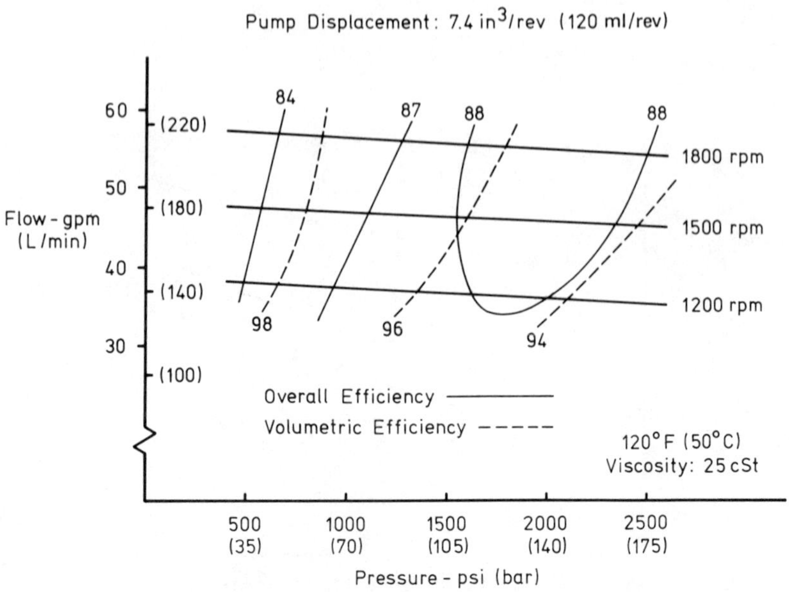

FIGURE 2-5 Industrial Balanced Vane Pump Performance

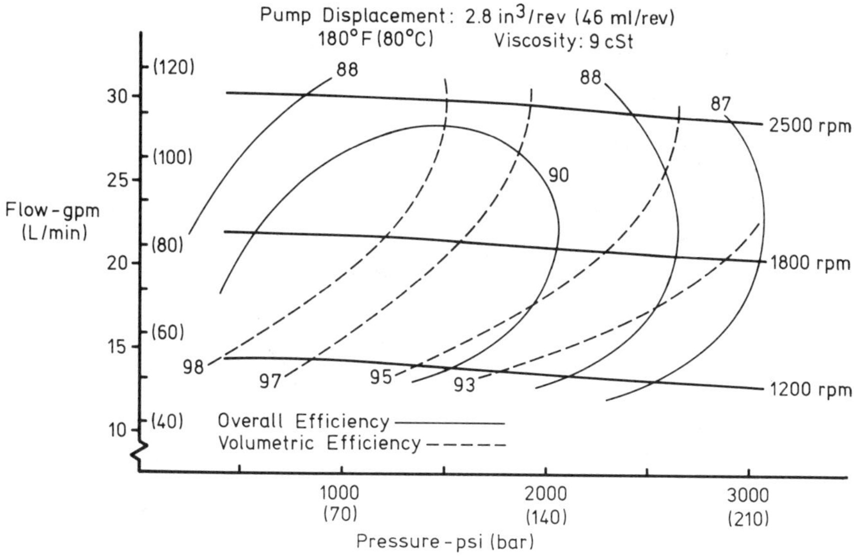

FIGURE 2-6 Mobile Balanced Vane Pump Performance

traction of the rotor. The plates are surfaced with a good bearing material for long life. The general construction is illustrated in Figure 2-7.

D. Balanced Vane Pump Cartridge Construction

Another feature of both the industrial and mobile units described above is their cartridge construction, illustrated in Figure 2-8. If reconditioning of the pump is required, it is accomplished by removing four bolts and replacing the pumping cartridge. The operation is simple and fast, and can often be done without disturbing the hydraulic connections.

E. Multiple Vane Pumps

Vane pumps are available in single, double and even triple units which have a common inlet and separate outlets, as shown in Figures 2-9, 2-10 and 2-11. Units with through-drive capability are also available, permitting combinations such as the fixed pump/variable pump assembly shown in Figure 2-12.

F. Offset Drives

Vane pumps are adaptable to the use of offset drives such as pulleys or gears. The models used for such applications will normally have an antifriction bearing

FIGURE 2-7 Balanced Vane Pump with Pressure-Loaded Flex Plates

FIGURE 2-8 Vane Pump Cartridge Construction

FIGURE 2-9 Single Vane Pump

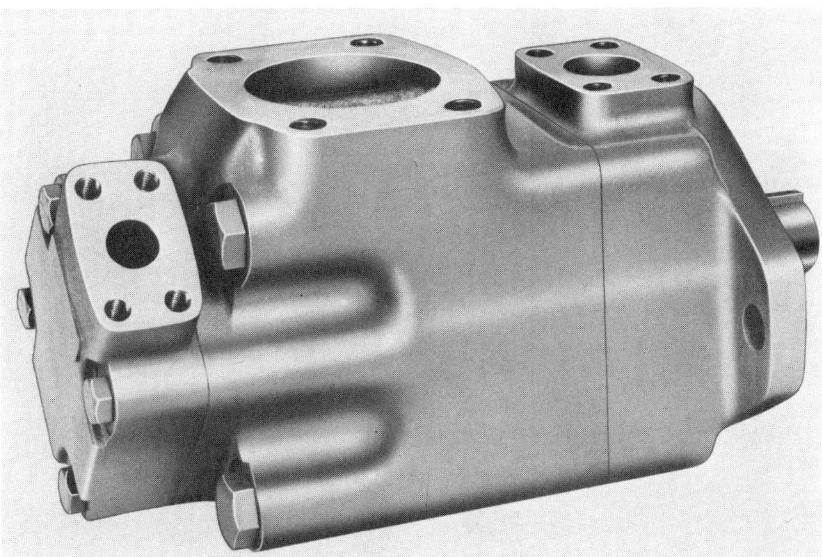

FIGURE 2-10 Double Vane Pump with Single Inlet

FIGURE 2-11 Triple Vane Pump with Single Inlet

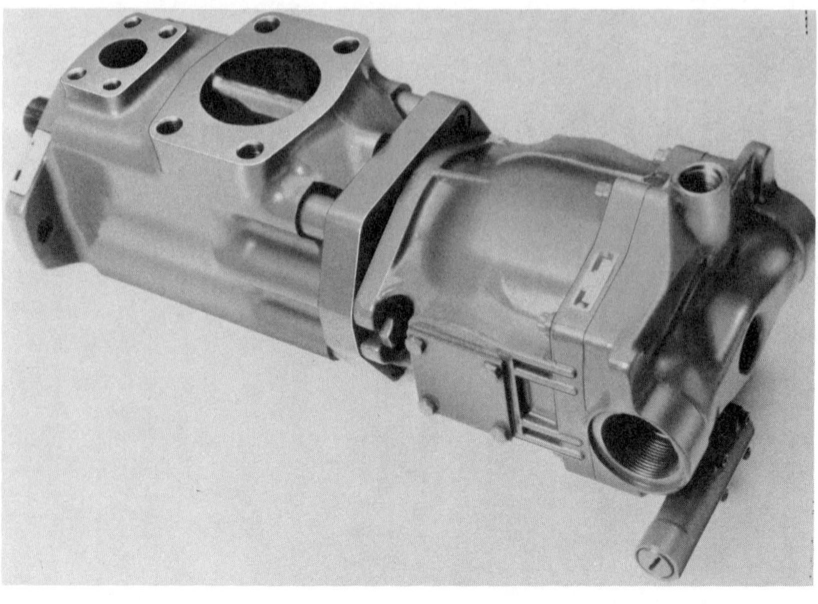

FIGURE 2-12 Vane Pump/Variable Piston Pump Assembly

at the drive end. Since the internal pumping forces place no side loads on the shaft bearing in a balanced vane pump, the full capacity of the bearing is available for such loads and the direction of the load application is immaterial.

IV. EXTERNAL VANE PUMPS

A. Operating Principle

Some vane pumps are made with a cam form on the outer surface of the rotor and the vanes are held in a stationary member which surrounds the rotor. This is essentially an inside-out version of the balanced vane pump previously described. When an even number of cam lobes is used, the pressure forces on the rotor are equal and opposite and there are no net shaft bearing loads.

The general operating characteristics of this type of design are very similar to those described for the balanced vane pump. Efficiency, durability and other factors are dependent on the detailed treatment of such design elements as vane loading, operating clearances and leakage paths—the same considerations that are present in other vane designs.

In general, sound levels are similar to those in the balanced vane design described in the previous section, but efficiency tends to be several percentage points lower. Manufacturers' literature should be consulted relative to any specific design.

V. EXTERNAL GEAR PUMPS

A. Configurations Available

External gear pumps have been widely used for many years in mobile and industrial hydraulics systems. They are simple, compact and adaptable to a broad range of applications. As with vane pumps, there are numerous configurations on the market including single pumps, double pumps with a common inlet and separate outlets, multiple pumps with individual inlets and outlets and through-drive pumps. These are shown in Figures 2-13 through 2-16.

B. Operating Principle

The gear pump produces pumping action by bringing oil from the inlet to the outlet in the spaces between the gear teeth and sealing off its return as the teeth mesh. Unlike the balanced vane pump which has no net shaft bearing loads, the gear pump bearings are positively loaded by the pressure forces on the gears as in Figure 2-17. These forces are substantial at normal operating pressures, making

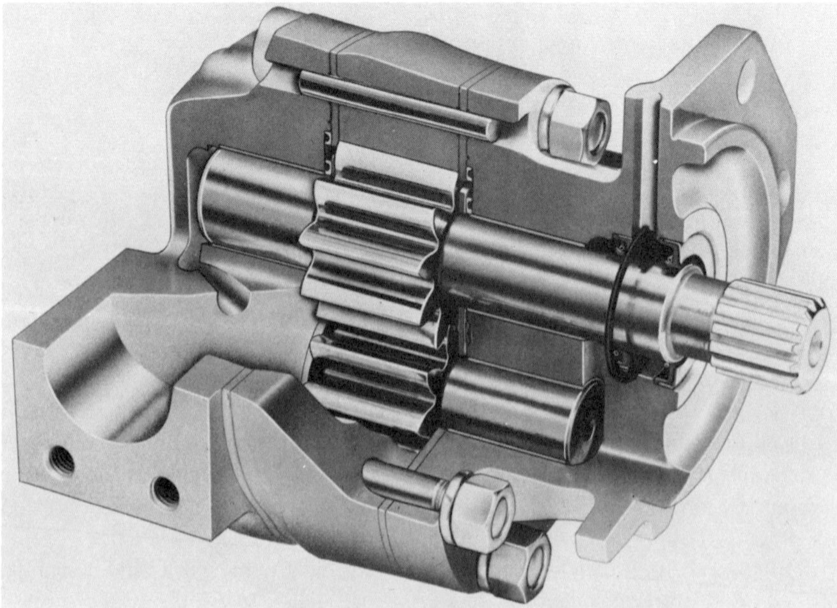

FIGURE 2-13 Single Gear Pump

FIGURE 2-14 Double Gear Pump with Single Inlet

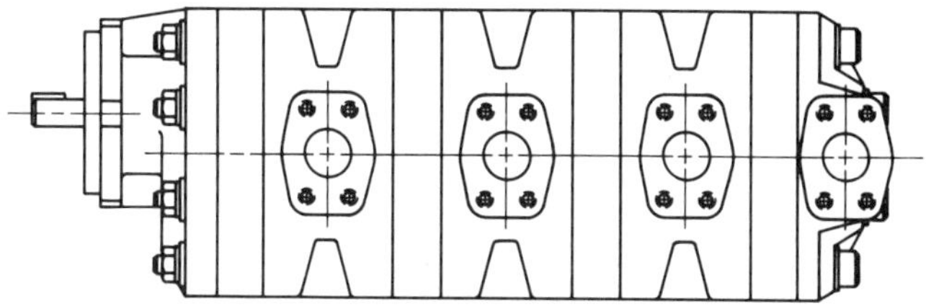

FIGURE 2-15 Quadruple Gear Pump with Multiple Inlets

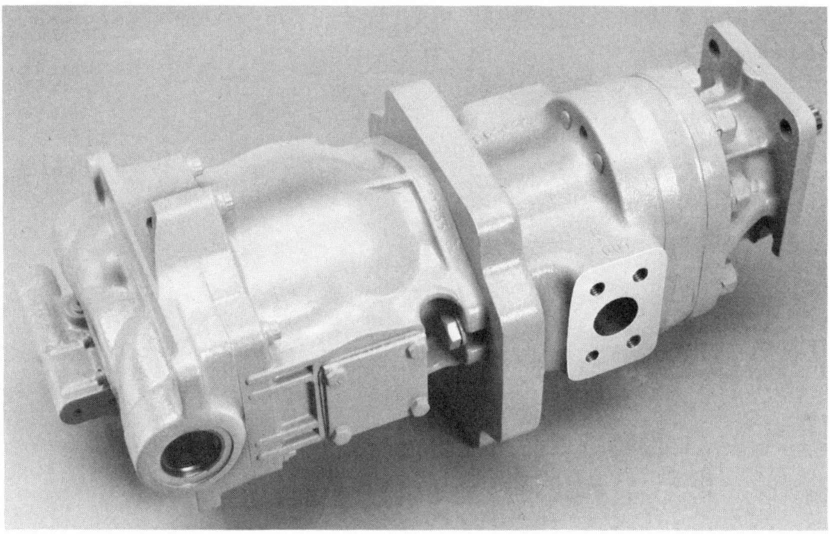

FIGURE 2-16 Through-Drive Gear Pump and Variable Piston Pump Assembly

the size and design of the bearings and the precision of their alignment critical factors in gear pump design.

As shown in Figure 2-18, the gear tooth contact forces add to the load on the driven-gear bearings but subtract from the driving gear bearing loads. The contact force is about 10% of the pressure force. This allows a spare capacity in the driving gear bearing to permit handling some offset load from a pulley or gear drive. It also means that, in cases of marginal operation when allowable bearing loads are exceeded, it is the driven bearing that can be expected to fail.

The gear-bearing loads are supported by hydrodynamic films when plain journal bearings are used. These require both a minimum rpm and a minimum fluid viscosity to provide an adequate film thickness to prevent metal to metal

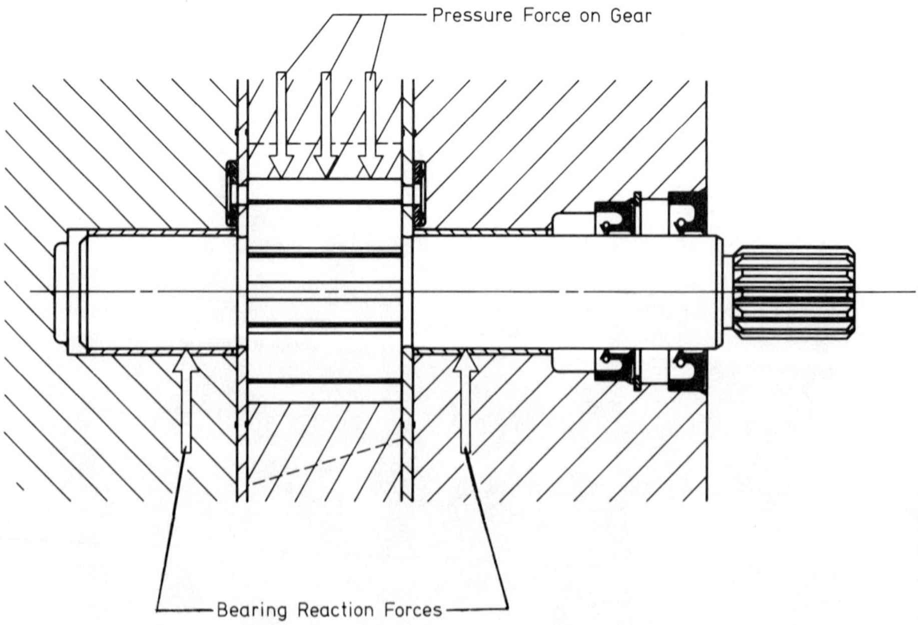

FIGURE 2-17 Gear Pump Pressure Forces

contact and wear. Operation at conditions involving high loads, low rpm and high temperature are therefore to be avoided.

1. Effect of Displacement on Bearing Loads On a gear pump that has gears of a given diameter, displacement can be increased by the use of wider gears. Since an increase in gear width exposes a greater area to pressure, the pressure rating of the pump must be reduced as gear width is increased to keep the gear-bearing loads constant. A family of gear pumps therefore customarily has ratings that can be charted as shown in Figure 2-19. For each frame size of pump, the combination of increasing displacement and declining pressure shown in the chart follows a constant torque curve: Pressure × displacement must be held constant to maintain constant bearing loads.

2. Pressure-Loaded Side Plates Most gear pump designs have some form of pressure-loaded side plates which bear against the faces of the gears as in Figure 2-20. These plates are needed to maintain high volumetric efficiency and to accommodate thermal changes that occur in the gears during operation.

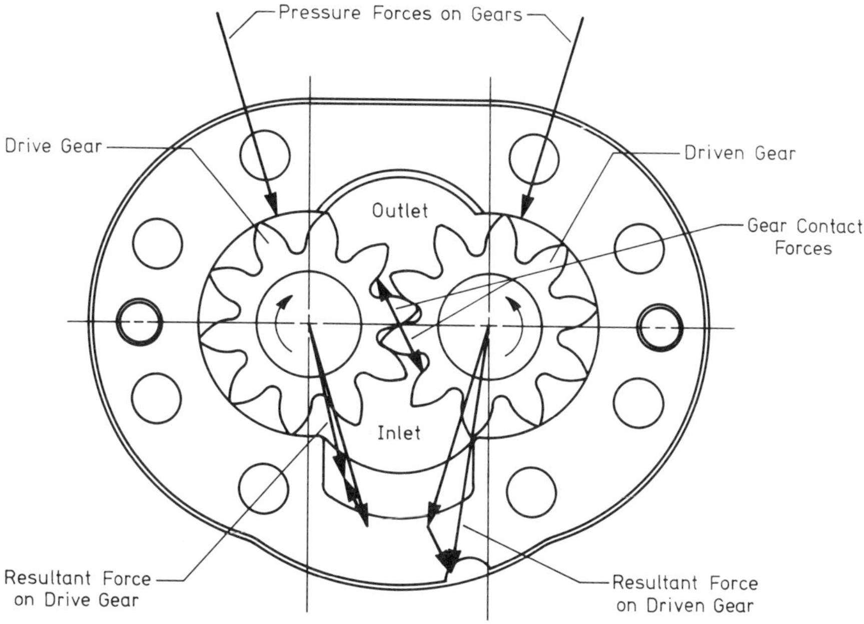

FIGURE 2-18 Effect of Gear Tooth Contact Forces*

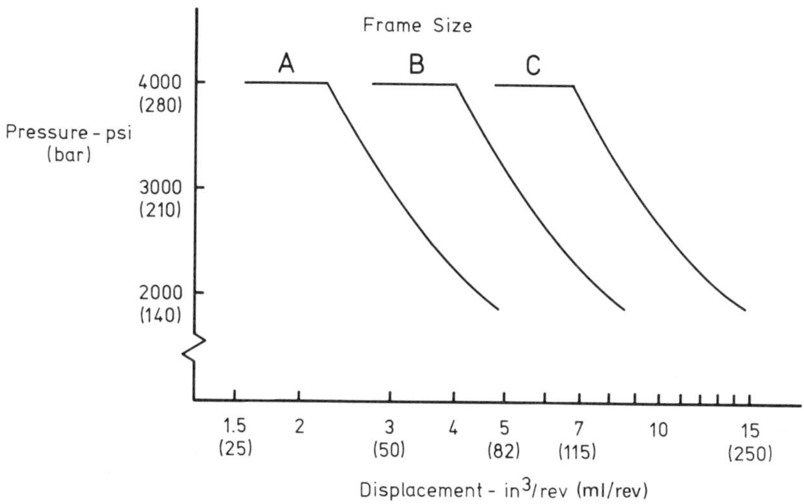

FIGURE 2-19 Gear Pump Family with Constant Bearing Loads

FIGURE 2-20 Gear Pump with Pressure-Loaded Side Plates

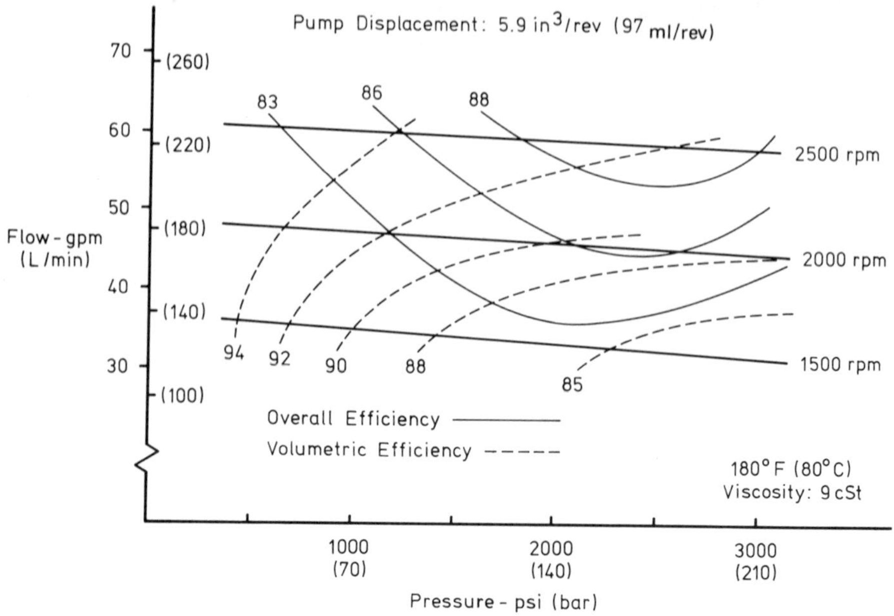

FIGURE 2-21 Medium-Duty Gear Pump Efficiency

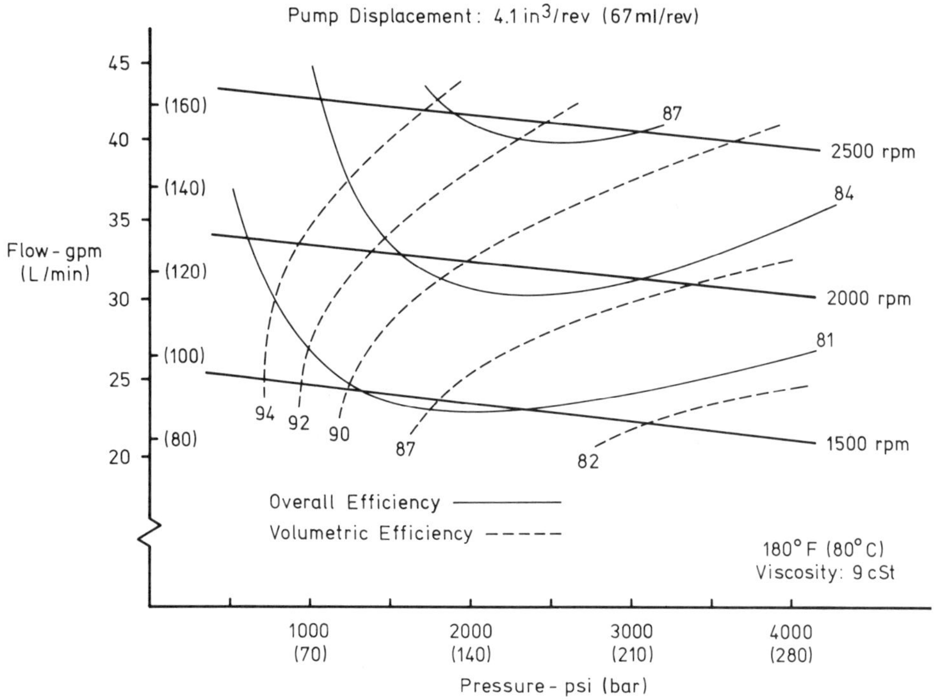

FIGURE 2-22 Heavy-Duty Gear Pump Efficiency

C. Gear Pump Performance

Typical performance curves are shown in Figure 2-21 and Figure 2-22. Pressure ratings falling between 2000 and 3000 psi (140 to 210 bar) are common for most designs and some heavy-duty pumps are rated as high as 4000 psi (280 bar).

D. Gear Pump Sound Levels

When care is taken in the design to avoid pulsing flow due to the trapping of oil between the meshing gear teeth, very favorable sound levels can be achieved. A sound chart for a typical heavy-duty pump operating at 3500 psi (240 bar) is shown in Figure 2-23. Sound level is 78 dB(A) or lower as measured by NFPA standard procedures. At 3500 psi (240 bar) and 2500 rpm, the pump shown is producing 80 hp (60 kW) of hydraulic power.

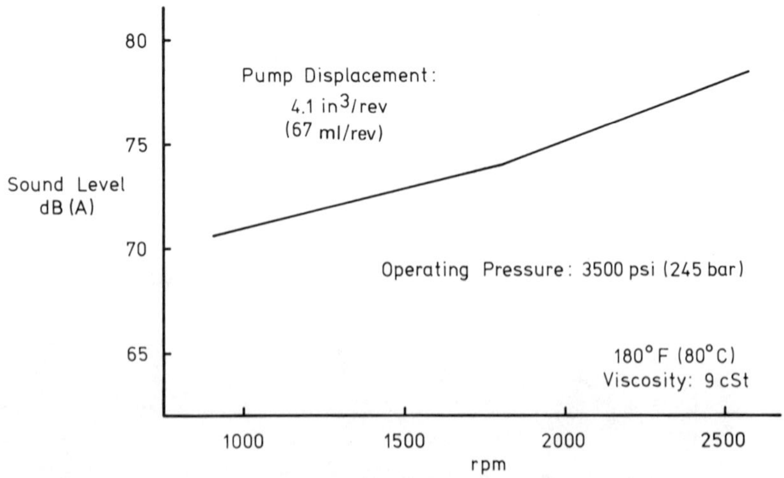

FIGURE 2-23 Gear Pump Sound Levels

VI. INTERNAL GEAR PUMPS

A. Introduction

Internal gear pump designs are simple and compact. They have an inherently low sound level. While operating pressure levels are generally not as high as in other fixed-displacement units, and efficiency tends to fall off as temperature increases, they are well suited to some industrial applications where the temperature is moderate, the loads are light or intermittent, and the total power consumption is low.

B. Operating Principle

The internal gear pump design consists of an inner and outer gear together with a crescent-shaped segment which acts as a seal at the gear tips. Figure 2-24 shows a typical configuration. As in the external gear pump, fluid is swept from inlet to outlet in the spaces between the teeth and its return to inlet is prevented by the meshing of the teeth.

In its simplest form, the gears rotate between rigid side plates with fixed clearance as shown in Figure 2-25. Some designs, however, have a pressure-loaded plate or plates to control the clearance at the gear faces and allow for thermal expansion of the gears during operation.

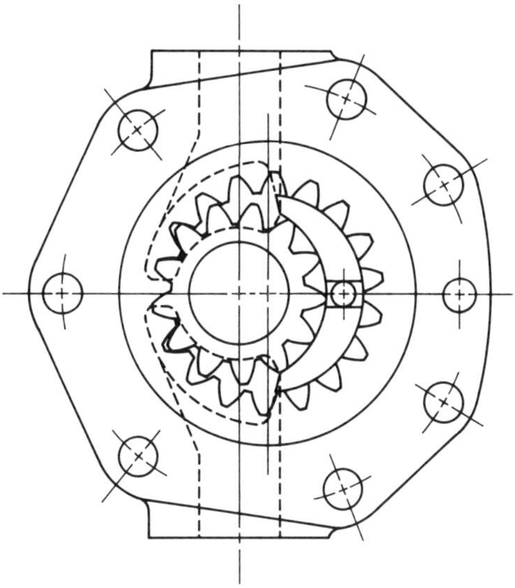

FIGURE 2-24 Internal Gear Pump

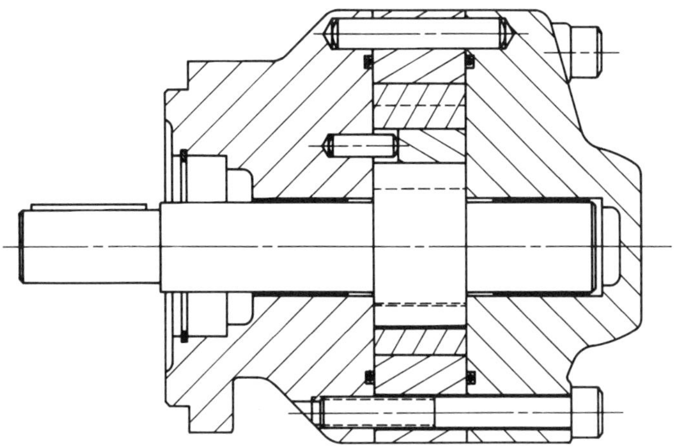

FIGURE 2-25 Internal Gear Pump with Fixed Side Plate Clearance

The radial pressure force on the inner gear is carried on journal bearings. As in the external gear pump, gear width may need to be reduced as pressure is raised to keep bearing loads within acceptable limits. The pressure force acting outward on the outer gear is at least partly counterbalanced by the hydrostatic forces exerted by the leakage fluid on the outer periphery of the gear.

C. Performance

While there are some internal gear pumps designed to operate at pressures above 3000 psi (210 bar), most are used in industrial applications at pressures of 2000 psi (140 bar) or less. The higher pressure designs usually employ not only pressure-loaded side plates, but also a crescent or wedge that is pressure loaded toward the gear teeth to control clearances and reduce leakage. Some form of positive hydrostatic support for the outer gear may also be provided.

An alternate means for achieving higher pressure output that is sometimes employed is to stage a multiple unit. The output of the first stage is ported directly to the inlet of the second stage, etc. In these designs, the pressure rise per stage is usually limited to about 1500 psi (100 bar). The multiple-stage design, of course, tends to negate the advantages of compactness and simplicity found in the simple single-stage pump.

The typical performance of a fixed clearance pump is shown in Figure 2-26. The use of features such as pressure-loaded side plates will help the pump to maintain reasonable efficiency as the temperature increases and fluid viscosity becomes less. An increase from the 100°F (38°C) temperature of Figure 2-26 to say, 160°F (70°C) would change viscosity by a factor of 3 or more, with leakage increasing accordingly in a fixed clearance unit.

The efficiency of internal gear pumps is good at lower temperatures and pressures. It can be expected to fall off somewhat as temperature increases and

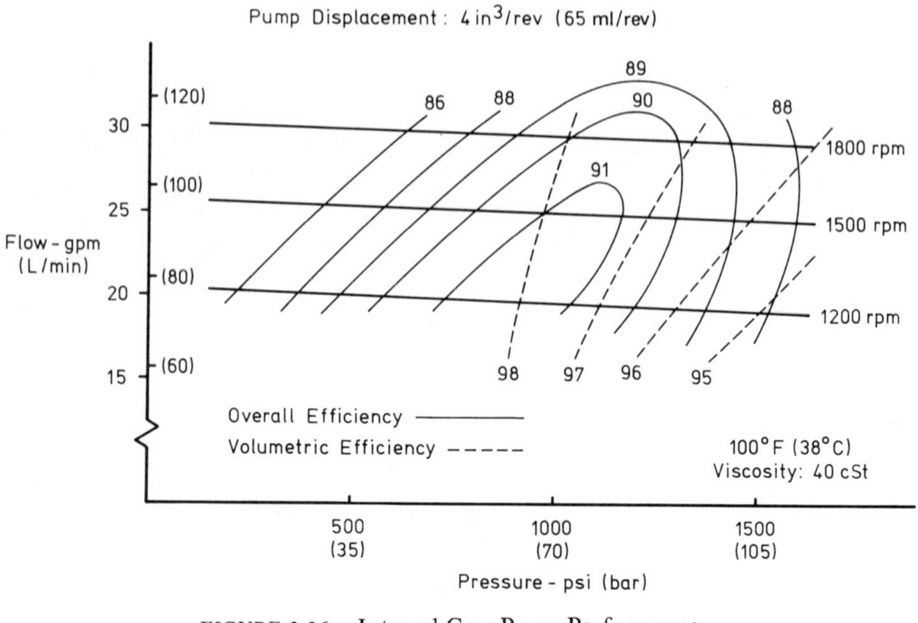

FIGURE 2-26 Internal Gear Pump Performance

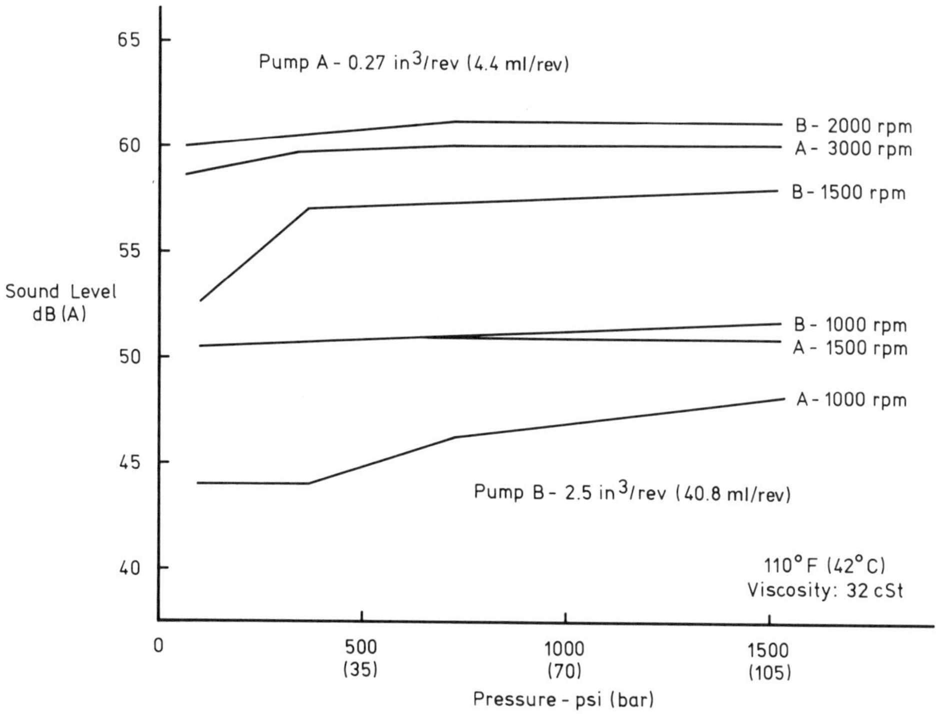

FIGURE 2-27 Internal Gear Pump Sound Levels

maximum operating conditions are approached. However, the offload power consumption is small and any performance penalty may be slight when the duty cycle is intermittent.

D. Sound Level

A very favorable characteristic of internal gear pumps is their low sound level. Figure 2-27 shows sound-level curves for typical units operating over a range of speeds and pressures. The sound level is uniformly low. The long arc of internal-external gear meshing provides a gradual tooth engagement and a smooth outlet flow, minimizing both fluidborne and airborne sound production.

E. Application Information

Internal gear pumps are compact in form and can be conveniently packaged in multiple units. Some models are available with built-in relief valves or other similar features.

Direct drive mounts are normally used because of limited side-load capacity in the internal pump journal bearings. When offset drives such as pulleys are employed, the pump is usually fitted with an additional antifriction bearing at the drive end to carry the added load.

The internal bearings within the pump depend on hydrostatic films and these are generated by the rotational velocity of the journals. At low speeds, the supporting film pressure may be inadequate if the pump is at full pressure. In addition, the net output flow of the pump falls off rapidly as speed is reduced and there is less flow through the pump to carry heat away. This internal temperature increase reduces fluid viscosity and also affects bearing capability.

The minimum speed for full pressure operation is usually in the range of 500 to 800 rpm. This poses no problem in electric motor driven units and is rarely a significant factor in any application if the pump can be unloaded at low speeds.

The inlet fill capabilities of internal gear pumps are very good and they will operate at normal speeds with atmospheric inlet or with a slight vacuum.

VII. FIXED DISPLACEMENT PISTON PUMPS

A. Operating Advantages of Piston Pumps

The fixed displacement pumps with the highest overall efficiency are of the piston type. Volumetric efficiencies above 97% and overall efficiencies around 90% are common. In addition, models are available that will give highly reliable, heavy-duty service at pressures of 5000 psi (350 bar) and sometimes more.

If the mode of operation is to include extended time periods of continuous full flow operation at pressures of 1500 psi (100 bar) or more, the energy saving due to the high efficiency of a piston pump can be substantial. Where power and/or pressure levels are high but the flow required by the load is variable, piston pumps are also generally used in a variable displacement version.

The higher efficiency of piston units is accompanied by higher cost, as compared with most other types of fixed pumps. Where power levels are low or duty cycles not severe, the energy saving may be insufficient to justify the added pump cost. In fact, for many circuits, pump efficiency may be overshadowed by power losses within the system itself, as explained in the section covering pump controls and systems.

B. Internal Construction

The two primary types of piston pumps in most widespread use are the inline piston design, shown in Figure 2-28, and the bent-axis design of Figure 2-29.

Both designs, in their fixed displacement versions, are used more frequently as motors. Both are also available in variable displacement versions. Piston motors and variable piston pumps will be discussed more fully in other chapters where their internal design features will be compared.

C. Piston Pump Efficiency

Typical efficiency curves for piston pumps designed to operate at typical medium-duty and heavy-duty pressure and temperature levels are shown in Figures 2-30 and 2-31. These curves are for inline units. Bent-axis pumps may run slightly higher but are at a cost disadvantage to inline units and are less frequently selected.

Of even greater importance than efficiency may be the ability of piston units to operate with high reliability at continuously high-power levels. The models designed for heavy duty are truly workhorse units capable of surviving pressure shocks, thermal excursions, and other severe demands without appreciable loss in life or performance.

This same ability makes piston units and particularly the inline design the only one capable of giving extended service on 5% soluble oil water based fluids at pressures of 2000 psi (140 bar) or higher. Other types of pumps are generally limited to 1500 psi (100 bar) or less. The high volumetric efficiency of inline pumps also gives them a decided advantage over other types in performance on such fluids. Even though leakage losses are much higher with fluids of very low viscosity, the piston pump suffers far less than the other types.

D. Piston Pump Sound Level

While generally not as quiet as vane units, the sound level of piston pumps can be brought down to very favorable levels by careful attention to internal design. The porting of fluid into and out of the pistons is of particular importance and the design can be "tuned" to be optimum over a regime of operation. Figure 2-32 illustrates the sound levels of three inline pumps, each designed to be optimum over the range of speeds and pressures commonly found in industrial applications.

E. Application Information

Both inline and bent-axis pumps have the drive shafts mounted on antifriction bearings. The capacity of these bearings is generally sufficient to allow operation with offset drives, such as pulleys or gears.

The high volumetric efficiency of piston units provides high overall efficiency down to relatively low speeds (typically 500 rpm or less) and there are no internal design features that limit full-load operation at low speeds.

Internal leakage of piston units flows into the pump case. This leakage could be ported back to the pump inlet but is more usually returned to the reservoir by means of an external drain line. Virtually all of the power-losses generated within the pump appear as a temperature rise in the case drain fluid.

During normal operating modes, including unloaded operation at moderate speeds, the internal leakage is high enough to carry away any losses generated within the pump case without excessive temperature rise in the oil. An exception may occur when a pump is run completely unloaded at high rpm (i.e., near its maximum limit). At low-outlet pressures, leakage to the case is extremely low and may well be insufficient to carry away the heat generated by the churning of oil within the case. While some heat will be lost by conduction to the pump mounting and by radiation or convection to the air, some supplemental cooling may be needed. This can be provided by bypassing a part of the return flow from the system open center directional valve through the pump case.

The internal construction of inline piston pumps includes a shaft passing through the center of the cylinder block. This makes it practical to design through-shaft versions in which an extension of the shaft can be used to drive a second unit.

FIGURE 2-28 Fixed Displacement Inline Piston Pump

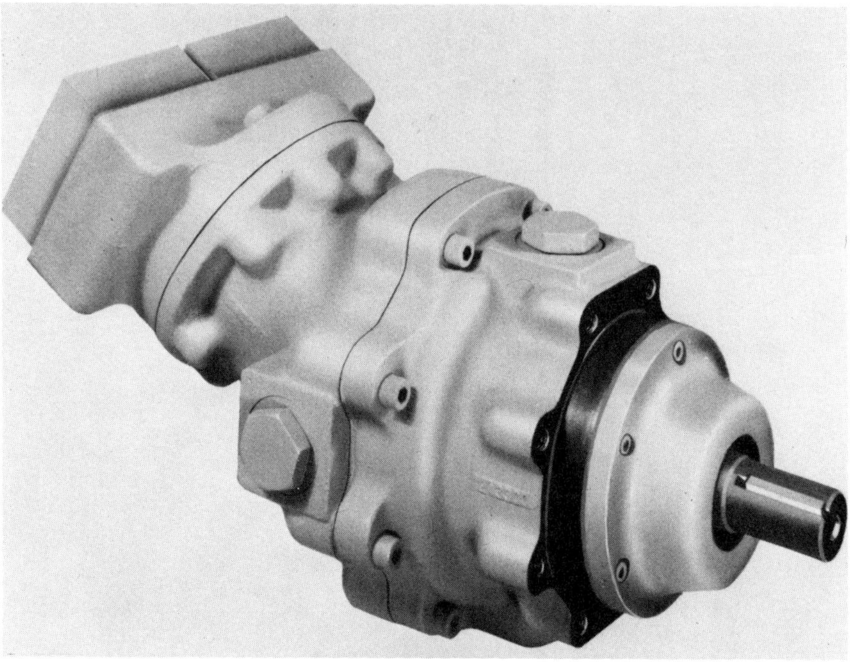

FIGURE 2-29 Fixed Displacement Bent-Axis Piston Pump

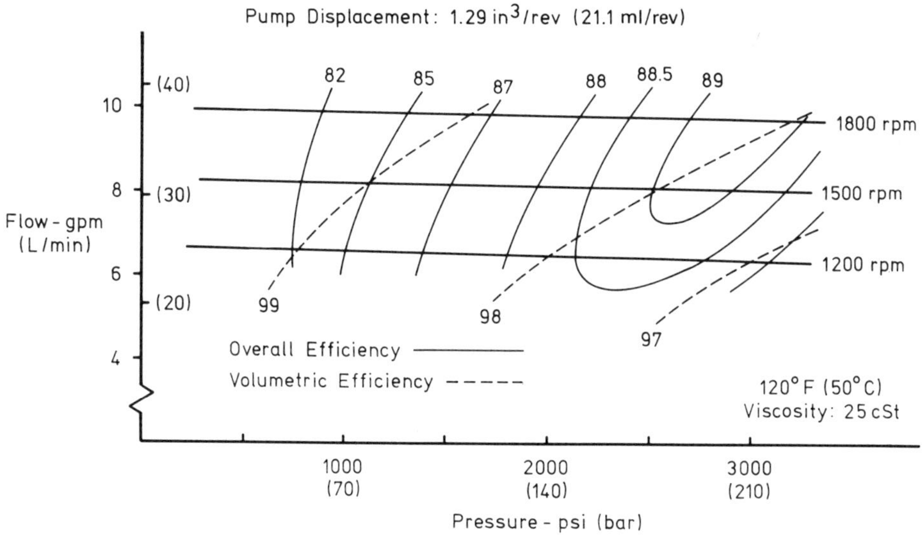

FIGURE 2-30 Medium-Duty Inline Piston Pump Performance

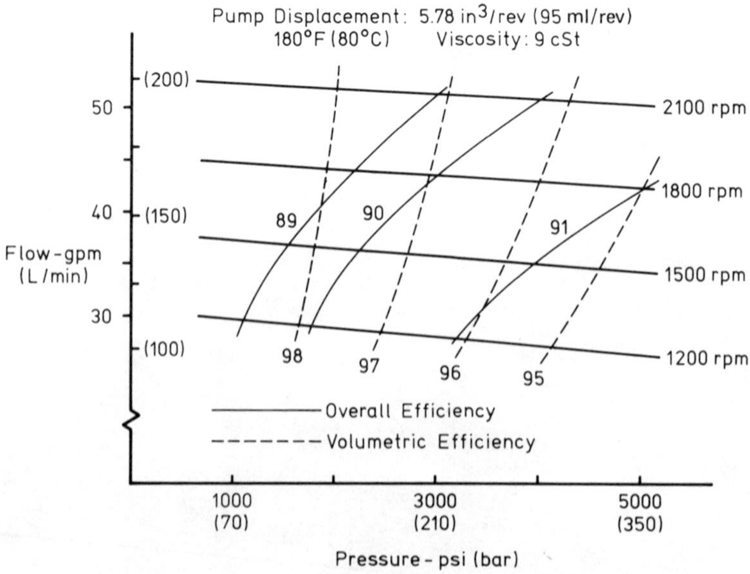

FIGURE 2-31 Heavy-Duty Inline Piston Pump Performance

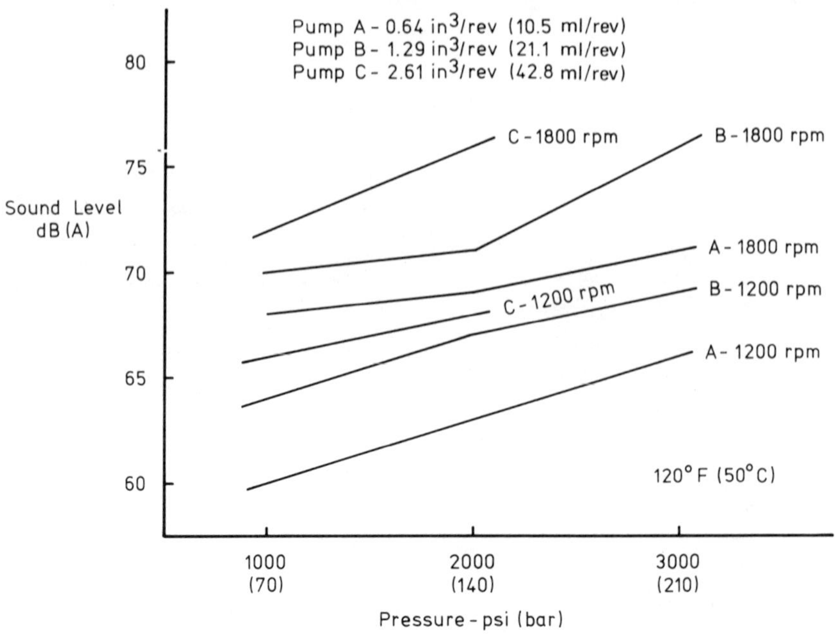

FIGURE 2-32 Industrial Inline Piston Pump Sound Levels

Typical applications of this type of unit include a large piston pump for primary vehicle power and a second smaller fixed-displacement unit for hydraulic power steering. In industrial applications, the small secondary unit sometimes provides the pressure and flow needed for servo controls.

3

Variable Displacement Hydraulic Pumps

I. INTRODUCTION

Hydraulic power provides a means for controlling motion smoothly and precisely. Metering the flow precisely provides both accurate positioning and excellent velocity control. The "stiffness" of a fluid column permits this accuracy to be maintained in spite of widely varying load conditions.

These functions of position and velocity control are achieved with the greatest energy efficiency by the use of variable displacement pumps. By proper application of the pump and its controls, the pump produces only the amount of flow required to move the load at the desired velocity. There is no excess of flow to be bypassed, as occurs when only a portion of a fixed displacement pump's flow is directed to the load.

The most commonly used types of variable displacement pumps are described in this chapter. The action of a typical pressure-limiting control or "pressure compensator" is also described. There are many other types of variable pump controls that are frequently used, either alone or in various combinations. These will be covered in Chapter 4: Pump Controls and Systems, where the various controls will be related to different system needs and the concept of "system efficiency" will be discussed.

II. INLINE PISTON PUMP

A. Operating Principle

The most common type of variable displacement pump in widespread use is the inline piston pump. It is simple, efficient, readily controllable and adaptable to a broad range of pressures, speeds, temperatures and fluids. The products of various manufacturers differ in numerous internal details, but all have certain features in common.

Figure 3-1 shows a cross-section of a typical inline piston pump. The pistons are contained in a rotating cylinder block which is mounted on a drive shaft. The ends of the pistons bear against a yoke or "swash plate" which produces the pumping motion. The pistons are returned, on the suction stroke, by a positive return mechanism.

The yoke can be varied in angular position. At its maximum angle, it produces the maximum stroking of the pistons and maximum flow. As the angle is varied toward zero, the flow diminishes to zero. If the angular motion continues beyond zero, pumping again begins but in the opposite direction. This feature is used to produce reversible flow operation in hydrostatic transmission pumps.

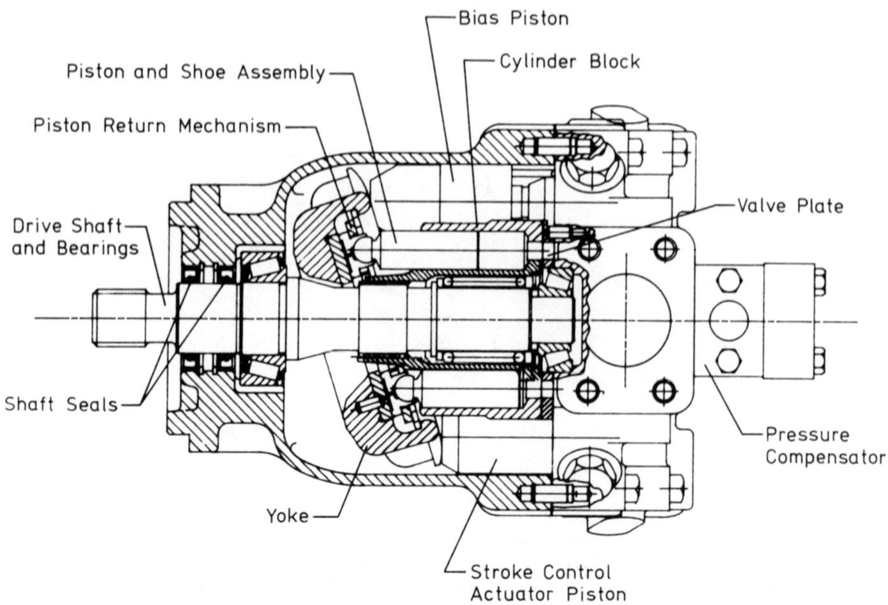

FIGURE 3-1 Variable Displacement Inline Piston Pump Cross-Section

FIGURE 3-2 Inline Piston-Pump Inlet and Outlet Porting

Fluid is metered into and out of the cylinder block through a valve plate or "port plate" as shown in Figure 3-2. The inlet and outlet passages are separated by valving lands which seal off the cylinder block ports as they pass from inlet to outlet and back at top and bottom dead center.

The primary piston pumping forces are carried on hydrostatically supported piston shoes, as indicated in Figure 3-3. Pump-outlet pressure is carried through the piston to a hydrostatic pool at the sliding surface of the shoe. The pressure in this pool and the leakage pressure gradient across the sealing surface carries nearly all of the piston pumping force, leaving only a small residual force which appears as a contact pressure on the sliding surface of the shoe and minimizes leakage from the hydrostatic pool.

A similar type of hydrostatic support is used at the face of the rotating cylinder block as also indicated in Figure 3-3. The leakage pressure gradients across the sealing lands largely balance the pressure forces inside the cylinder bores. The small remaining hydrostatic unbalance, together with a spring force, ensure positive seating of the cylinder block against the valve plate for minimum leakage. These seating forces are large enough to resist the cylinder block tilting moment caused by centrifugal force acting on the pistons, as illustrated in Figure 3-4. The forces are usually carried not only on the cylinder block sealing

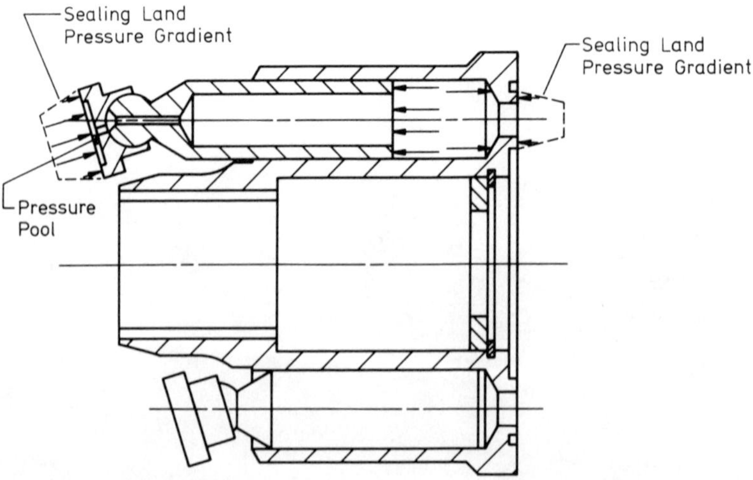

FIGURE 3-3 Hydrostatic Support of Cylinder Block and Piston Shoe

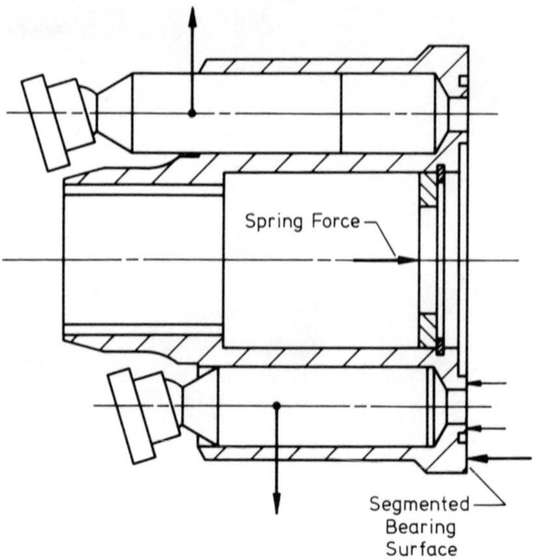

FIGURE 3-4 Centrifugal Tilting Moment of Inline Pump Pistons

lands but also on a segmented bearing surface at the outer edge of the cylinder block.

Leakage past the pistons, piston shoes and cylinder block face is carried through the pump housing and out a case drain connection to the reservoir. This leakage flow carries away the heat generated by the volumetric and mechanical inefficiencies of the pump and prevents any heat buildup and overheating within the pump.

B. Automatic Stroke Control

Variable displacement pumps are commonly fitted with a pressure-sensitive automatic stroke control, usually called a "pressure compensator" or "pressure limiting control." A typical construction is shown in Figure 3-5. Pump-outlet pressure is applied to a spring-loaded spool valve. When the pump-outlet pressure exceeds the spring setting, fluid is metered to the pump stroke control actuator to reduce pump outlet flow, causing the pump-outlet pressure to drop. If pump-outlet pressure is too low, fluid is metered from the stroke control actuator, causing pump-outlet flow to increase and the pressure to rise until a balance is again restored.

The pump is initially positioned in positive displacement by a spring. The design shown is further urged in that direction by a bias piston, located opposite the control piston and connected to outlet pressure. The actuator piston is larger in diameter than the bias piston and can override it to reduce the pump stroke.

This type of pressure compensator and stroke control mechanism will maintain nearly constant pressure at the pump outlet throughout the entire flow range of the pump. The maximum pressure rise from full flow to zero flow is usually 3% to 5%. Figure 3-6 is a typical pressure-compensated pump flow curve. When operated in a closed center system as described in Chapter 4, the pump will normally operate at some point along the vertical portion of the curve, delivering a controlled flow at full pressure. The pump will reach the horizontal portion of the curve only if the system valve is fully open and the pressure required by the load is low.

Since the distance to be traveled by the stroke control actuator is short and the inertial forces of the yoke and pistons are relatively small, response times

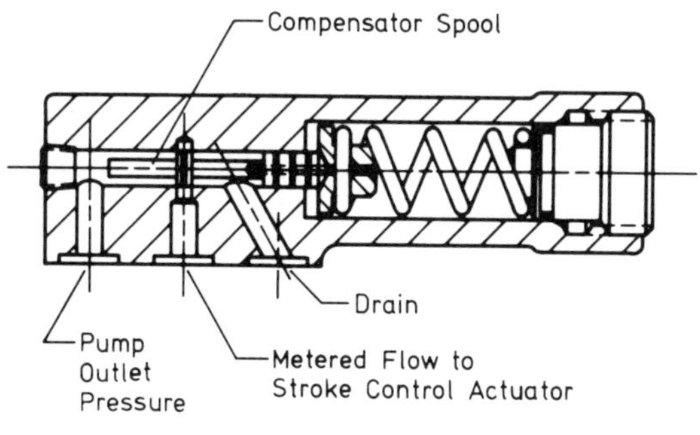

FIGURE 3-5 Variable Displacement Pump Pressure-Compensator Control

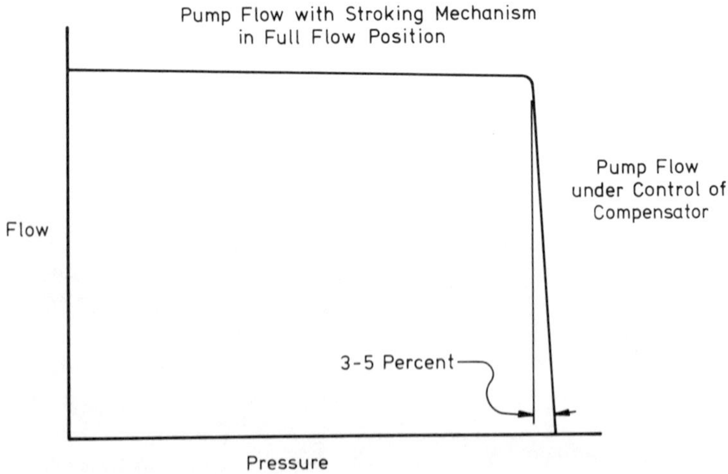

FIGURE 3-6 Pressure-Compensated Pump Flow Curve

can be very rapid. The time required for the yoke to move from the full stroke position to zero to control a pressure increase is typically about 20 to 40 milliseconds.

C. Pressure and Speed Ratings

In the inline design, the hydraulic pressure forces on the cylinder block and piston shoes are largely hydrostatically balanced. In addition, the piston side forces which produce loads on the rotating shaft bearings virtually disappear when the yoke is positioned for low flows. A large percentage of a variable pump's operating time is usually in the low-flow range. These features, hydrostatic pressure balance and low shaft-bearing loads, allow inline piston pumps to be designed for long life at high operating pressures. Pressure ratings of 5000 psi (350 bar) or more are achieved by some designs.

The speed rating of all pumps is limited by inlet-pressure considerations. The pumping action of any type of positive displacement pump creates a partial vacuum in the pumping chamber and the absolute pressure in the inlet fluid must be sufficient to accelerate the fluid to the velocity of the vanes, gear teeth or pistons to prevent cavitation damage. Piston pumps are slightly more limited in this regard than are some of the more compact, fixed-displacement units. They are, nevertheless, capable of operating at very acceptable speeds at atmospheric inlet pressure and even at slight inlet vacuums.

However, the speed range of piston pumps may be extended further by supercharging the inlet flow. One means is through the use of a pressurized

FIGURE 3-7 Inline Piston Pump with Integral Full-Flow Supercharge Pump

reservoir as is common in aircraft pump circuits. Alternately, a separate supercharge pump is sometimes used. Figure 3-7 shows such a combination with the supercharge pump being an integral part of the package and driven by the same shaft.

The supercharge pump allows an increase of 30% or more in the speed rating of this family of piston pumps, yielding considerably higher power from the same package. It also has other important advantages:

The flow from the supercharge pump can be filtered before entering the main pump. This not only increases the life and reliability of the main pump, but of the other downstream-system elements as well.

The excess of flow from the supercharge pump can be passed through the case of the main pump for added cooling.

The fluid in the inlet line from the reservoir is in constant motion and does not need to be accelerated when a sudden flow demand is placed on the main pump. This improves the speed of system response and avoids periods of momentary cavitation at the main pump inlet.

Chapter 6 discusses hydrostatic transmissions operating in closed loop circuits. In that type of circuit, a secondary pump is used to keep the closed loop supercharged. Like the full-flow supercharge pump mentioned above, it allows the main transmission pump to operate at higher speeds without cavitation. It also allows filtration of the fluid entering the transmission circuit and provides a means of auxiliary cooling flow for the transmission pump.

D. Efficiency

The efficiency of the inline pump is excellent at full flow and power loss diminishes as flow is reduced. Because of their high volumetric efficiency, piston pumps are less affected by viscosity reductions at higher temperatures than other pump types. The result is a reasonably constant efficiency over a broad band of temperature.

1. Medium-Duty Piston Pump Figure 3-8 shows typical efficiency values for an inline piston pump over a range of pressures and speeds. These are the full flow or "corner horsepower" efficiency regimes. The efficiency is 90% or above over the greater part of the operating range.

The effect of temperature on volumetric efficiency for this pump is il-

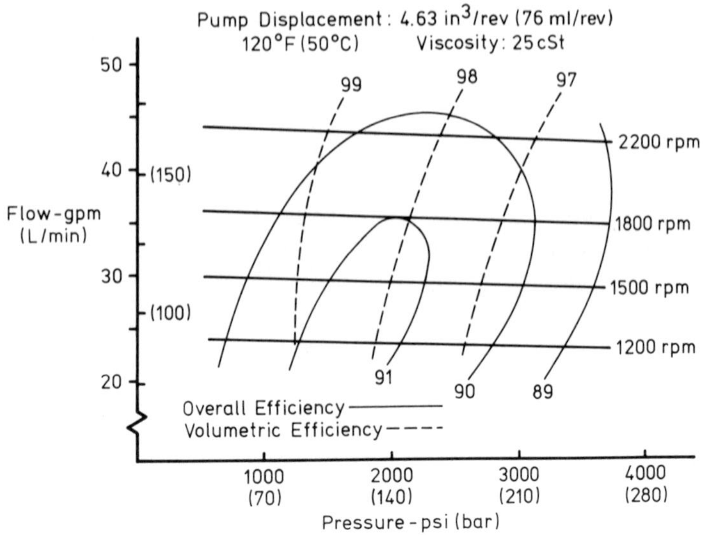

FIGURE 3-8 Medium-Duty Inline Piston-Pump Efficiency for Full-Flow Operation

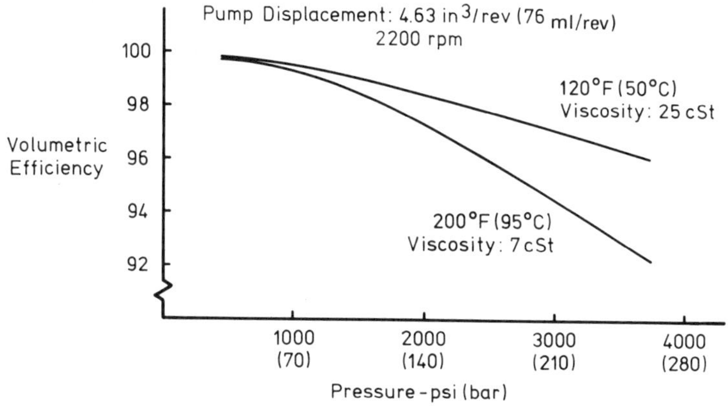

FIGURE 3-9 Inline Piston-Pump Volumetric Efficiency Change with Temperature

lustrated in Figure 3-9. The reduction is less than 3.5% at 3625 psi (250 bar) in going from 120° to 200°F (50° to 95°C) even though the viscosity changes by a factor of more than 3.

In the inline piston type of variable displacement pump, the full-flow efficiency is high and the power loss becomes less as pump stroke is reduced at constant pressure. This is because certain of the sources of loss—flow losses through the pump passages and piston friction, for example—diminish as the stroke goes toward zero. A typical curve of power loss versus flow is shown in Figure 3-10. At the zero-flow condition, if a load-sensing compensator is used as described in Chapter 4, the pump can be depressurized to a low-pressure level, reducing pump internal leakage. The resulting standby power level is only a fraction of that at full pressure and zero flow.

2. *Heavy-Duty Piston Pump* As previously noted, heavy duty pumps designed for operation in the 5000 psi (350 bar) range are sometimes equipped with full-flow supercharge pumps. This permits operation at higher rotational speeds without cavitation.

Typical efficiency values for a supercharged heavy duty pump operating at 180°F (80°C) are given in Figure 3-11. These are the full flow or corner horsepower efficiency values. The high volumetric efficiency of the inline design allows overall efficiency to remain very favorable in spite of the elevated pressure-level and low-fluid viscosity. Power loss will diminish as flow is reduced, as with the medium-duty pump, and depressurization at zero flow offers a further opportunity for power saving.

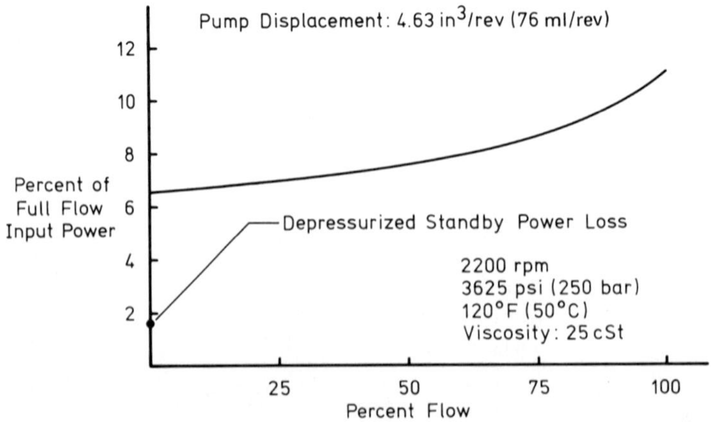

FIGURE 3-10 Inline Piston-Pump Power Loss During Part-Flow Operation

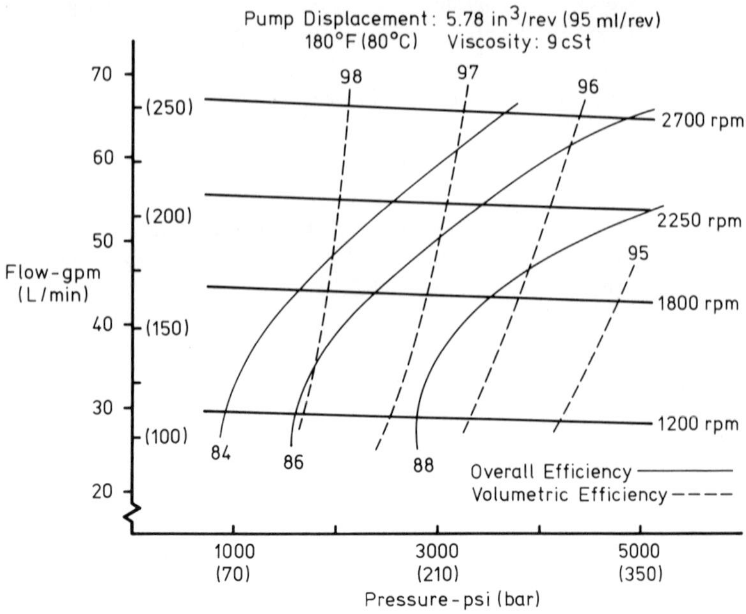

FIGURE 3-11 Heavy-Duty Inline Piston-Pump Efficiency for Full-Flow Operation

3. Industrial Piston Pump Industrial piston pumps are expected to operate efficiently at electric motor drive speeds between 1200 and 1800 rpm and over the pressure range from 1000 to 3000 psi (70 to 210 bar). Figure 3-12 shows the pattern of full-flow efficiencies over the range of speeds and pressures for a typical industrial pump.

Again, as with medium- and heavy-duty designs, power loss will diminish at partial flows. Also, standby power loss can be further reduced by depressurization using a load-sensing control.

E. Sound Level

The sound level of piston pumps can be minimized by careful attention to the porting of fluid to and from the pistons. The lowest sound levels are achieved when the change of pressure within each piston cavity is smooth and steady with no abrupt changes or shocks.

This can be achieved by setting up a computer simulation of the pressure at each piston, taking into account the operating speed, the pressure level, and the compressibility of the fluid. The pressure profile is then smoothed out by varying the width of the sealing land between the inlet and outlet, and by adjusting the size and position of metering grooves. When properly done, this analysis will not only result in the lowest possible sound level but will also maximize pump life by avoiding shock loads or high, local-fluid velocities that could produce erosion.

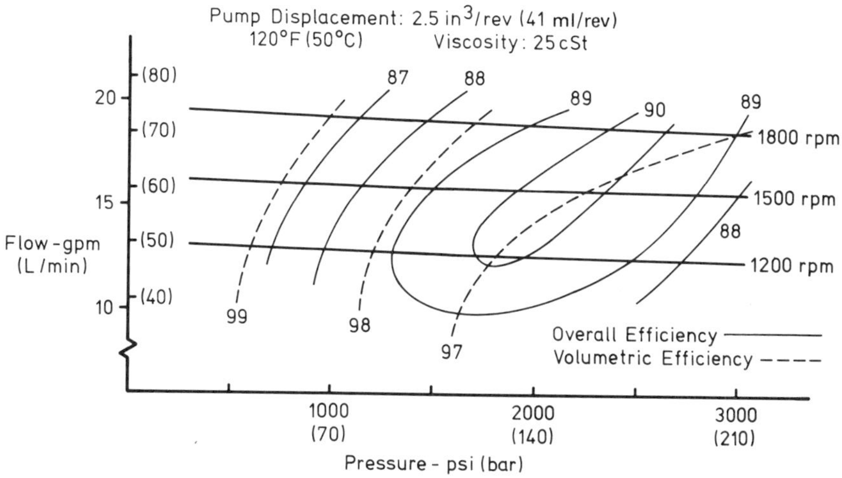

FIGURE 3-12 Industrial Inline Piston-Pump Efficiency for Full-Flow Operation

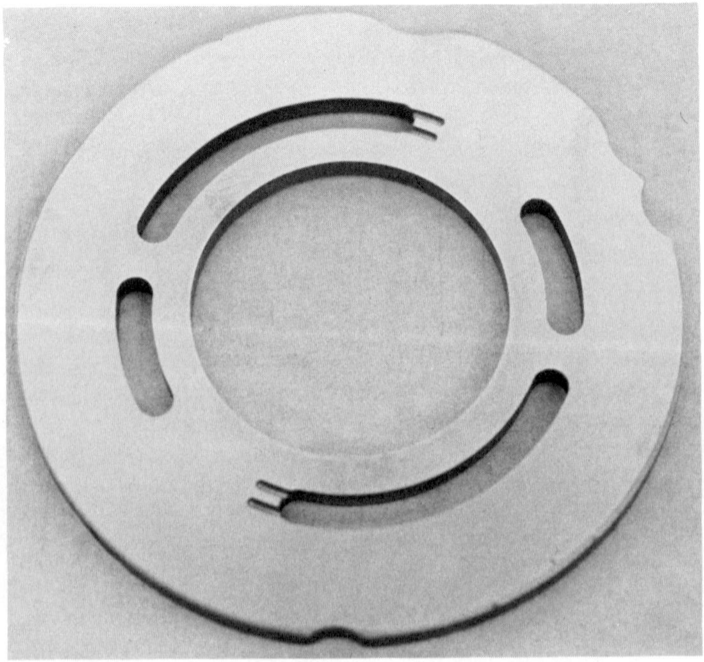

FIGURE 3-13 Piston-Pump Valve Plate with Metering Grooves for Sound Reduction

A typical piston-pump valve plate is shown in Figure 3-13. In this case, twin metering grooves of rectangular cross-section are used. The computer study assures that the design will be optimized over the full range of expected speeds and pressures. This is later verified by actual test.

Figure 3-14 shows the sound levels achieved in two versions of the design covered in Section 3. One has been optimized for pressures to 3000 psi (210 bar) and speeds to 1800 rpm. It has a sound level below 80 dB(A) at 1800 rpm and below 75 dB(A) at 1200 rpm. The second version was specifically optimized for low-pressure, low-speed machine tool applications. Its sound level is below 65 dB(A).

The possibility of optimization for minimum sound level is not confined to the pressure and speed regimes of industrial units. Similar treatment has been applied in specific applications of hydrostatic transmission pumps which operate at speeds of 2400 to 3600 rpm and pressures up to 5000 psi (350 bar). By studying the transmission duty cycle and optimizing the design for the most frequent operating modes, savings of 5 to 8 dB(A) have been achieved.

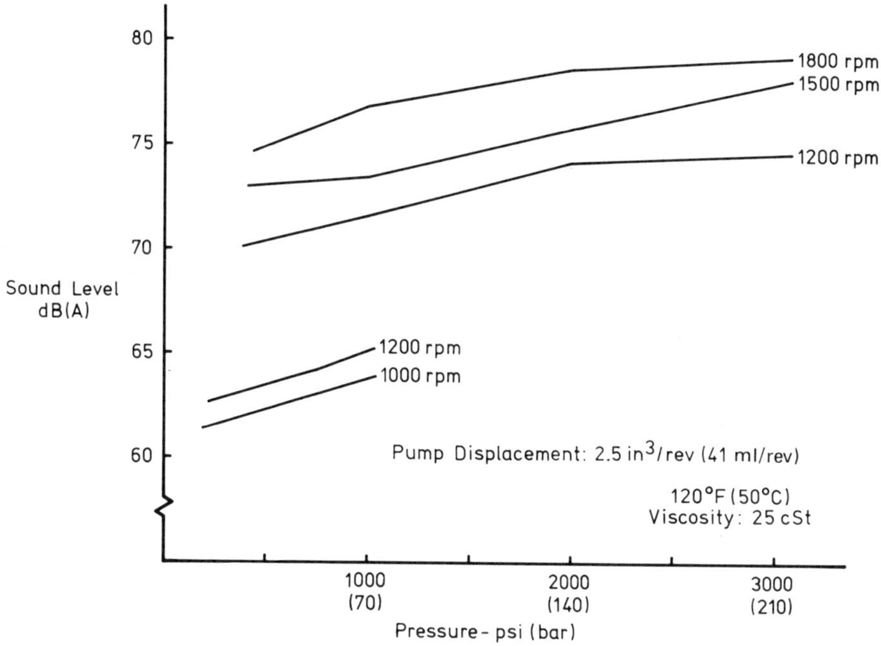

FIGURE 3-14 Sound Levels of Inline Pumps Optimized for Industrial Applications

III. BENT-AXIS PISTON PUMP

A. Operating Principle

A second type of high-performance, variable piston pump is the bent-axis design. Like the inline, it can be designed to provide long life under heavy-duty conditions. While efficiency may be equal to or slightly higher than that of comparable inline units, the design is more expensive to produce and is less frequently used than the inline.

In the bent-axis design, the cylinder block is at an angle to the drive shaft and the pistons are joined to the drive shaft with connecting rods, as shown in Figure 3-15. Rotation of the drive shaft, which is mounted on anti-friction bearings, provides pumping action. As the axis of the cylinder block is swung to bring it into alignment with the drive shaft, pumping is reduced to zero.

The angularity of the connecting rods with the cylinder bores is very small which minimizes any radial side forces between the pistons and the cylinder walls. The cylinder block rotation is synchronized with the drive shaft by a constant-velocity, universal link. The universal link carries none of the pumping power, since the connecting rods are held in alignment with the cylinder bores.

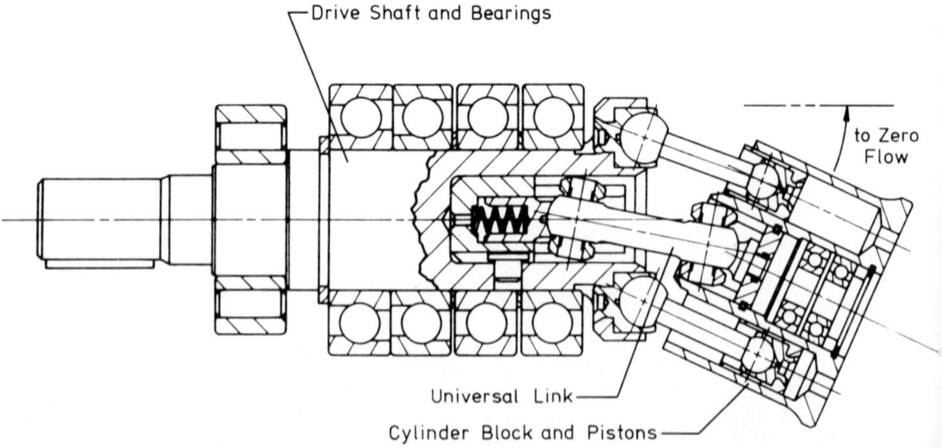

FIGURE 3-15 Bent-Axis Pump Cylinder Block and Drive-Shaft Construction

The link need only supply the torque needed to satisfy the rotational friction and windage losses of the rotor.

Valving of the fluid into and out of the cylinder block is through a valve plate as in the inline pump. The entire cylinder block must be swung through the axis angle to achieve variable delivery, while still maintaining hydraulic connections to the inlet and outlet ports of the external pump housing. One means of accomplishing this is through a swinging yoke mechanism that carries the cylinder block and allows its angle to be varied. The fluid is carried in and out of the yoke through swivel joints at the yoke swing axis. Figure 3-16 shows a complete bent-axis pump assembly. One of the yoke port connections, covered by a shipping cover is visible at the right side of the assembly.

B. Stroke Control

The same types of automatic stroke control mechanisms are applicable to the bent-axis pump and to the inline design. The only difference is that the entire mass of the cylinder block and the surrounding porting mechanism must be moved through a considerable distance in the bent axis design. The inertial effects result in slower response than in the inline units. The times are typically 5 to 10 times as long and must be allowed for in the system design.

C. Application Characteristics

Speed ratings for unsupercharged bent-axis and inline units are comparable, although it should be noted that the bent-axis design does not lend itself to the

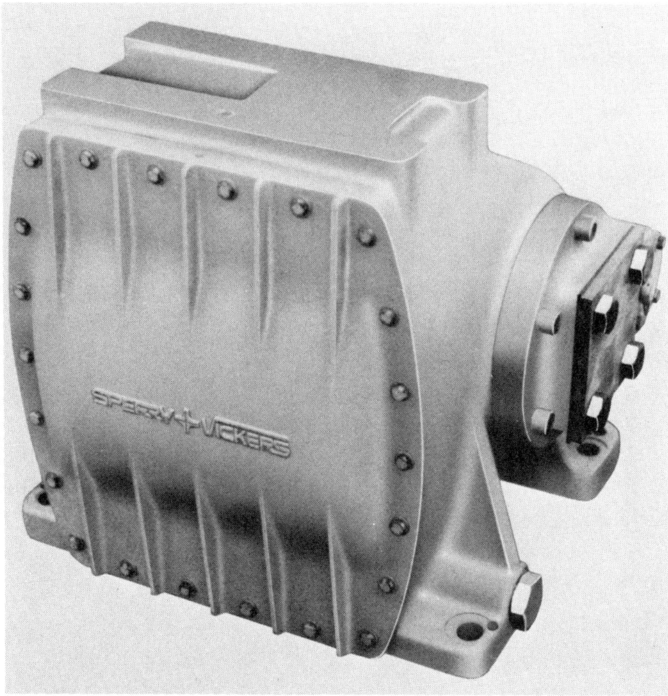

FIGURE 3-16 Bent-Axis Pump Assembly

addition of an integral supercharge pump for extension of the speed range. The pressure rating of the bent-axis design is largely dependent on the capabilities of the antifriction bearings chosen to support the drive shaft. These bearings remain constantly loaded as long as the pump is pressurized, regardless of stroke position. Bearing B-10 life is usually the limiting factor in pump-service life in the medium and heavy-duty applications.

Efficiency of the bent-axis pump is equal to or sometimes slightly higher than the inline. Also, similarly, power loss reduces as displacement is reduced. Depressurized operation for minimum standby power loss is also possible.

As with the inline pump, there is a separate case drain connection for carrying leakage fluid back to the reservoir, assuring internal cooling of the pump.

The sound level of bent-axis pumps is dependent on the care taken in designing the port-timing, as with other pump types. The design has been successfully applied in a variety of industrial and mobile systems without limitations due to sound levels or other basic-rating characteristics.

IV. SINGLE ECCENTRIC VARIABLE VANE PUMP

A. Operating Principle

The most common type of variable vane pump is the single eccentric type. It has found wide use in industrial applications because of its low sound levels and compactness. This type of pump can be easily stacked to form multiple units. Its efficiency is not as high as that of variable displacement piston pumps, but when properly applied, it does offer significant power savings over fixed displacement designs in many industrial circuits.

In the typical construction shown in Figure 3-17, the rotor and vanes revolve inside a circular cam ring. Fluid is ported into the spaces between the vanes as they are moving outward and is pumped out the outlet passage as the vanes move inward. Valve-plate construction, similar to that in piston pumps, has the inlet and outlet ports separated by sealing lands.

The eccentricity of the cam ring can be varied to control the motion of the vanes. As the ring is moved toward the centered position, vane motion is reduced and the displacement of the pump moves toward zero. Cam position is normally controlled by a pressure compensator which meters oil to and from an actuator piston as in the variable piston pumps.

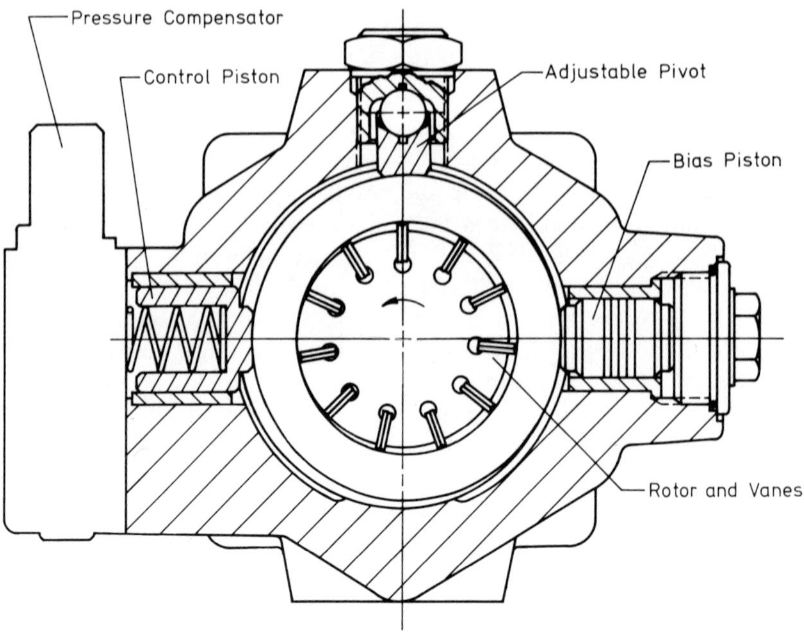

FIGURE 3-17 Single Eccentric Variable-Vane Pump Cross-Section

In the design shown, the cam ring is placed in its initial off-center position by a spring and further aided by pump-outlet pressure metered to the control actuator piston by the pressure compensator. The cam ring is urged toward its centered position by a smaller bias piston that is continuously connected to pump outlet pressure. In this design, the pump displacement is moved toward zero flow by metering fluid out of the actuator piston, rather than into it as in the piston pump of Figure 3-1. The overall functioning of the pressure compensator and stroke control mechanism are unaffected by this reversed sense and the same accuracy of control is achieved.

The hydraulic pressure reaction force on the inner surface of the cam ring is carried by a pivot in the main housing of the pump. Adjustment of this pivot varies the amount of fluid compression and decompression that occurs as the vanes pass across the valving lands between the inlet and outlet ports. This permits adjustment to be made to suit the primary operating pressure level of the application, resulting in minimum sound level. There is also a small but beneficial effect on efficiency.

The reaction force of the outlet pressure which acts over 180° of the rotor periphery is carried on plain bearings, as shown in Figure 3-18. As in the case of

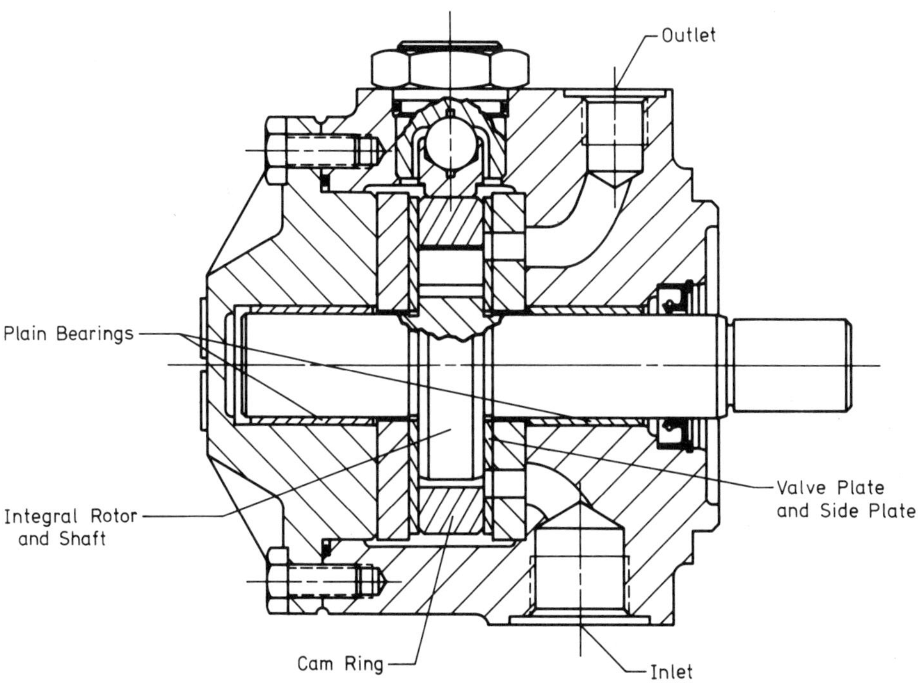

FIGURE 3-18 Single Eccentric Variable-Vane Pump Bearing and Side-Plate Construction

fixed gear pumps described in Chapter 2, the width of the variable vane rotor must be adjusted as the pressure rating is increased to keep bearing loads within allowable limits. The design of the bearings and the precision of alignment are key factors in achieving reliable operation at rated pressures.

The rotor and vanes rotate between side plates, one of which is used to valve the fluid from inlet to outlet. Running clearances must be controlled to minimize leakage past the rotor and vanes and also past the cam ring, which must be free to move to adjust pump displacement.

B. Performance

The single eccentric variable-vane pump finds its best performance over the range of speeds, pressures and temperatures characteristic of industrial applications, e.g., speeds corresponding to electric motor-drive speeds from 1000 to 1800 rpm and temperatures in the 120°F (50°C) range. Pressure is generally limited by bearing loads. The smaller units having displacements up to about 5 in^3/rev (80 ml/rev) are rated up to 2500 psi (175 bar). Larger units, up to about 10 in^3/rev (160 ml/rev) are rated in the 1000 to 2000 psi (70 to 140 bar) range.

Full-flow overall efficiency of this type of pump is typically in the vicinity of 80% to 87% over the range of industrial operating conditions. Figure 3-19 is characteristic of the performance that can be expected.

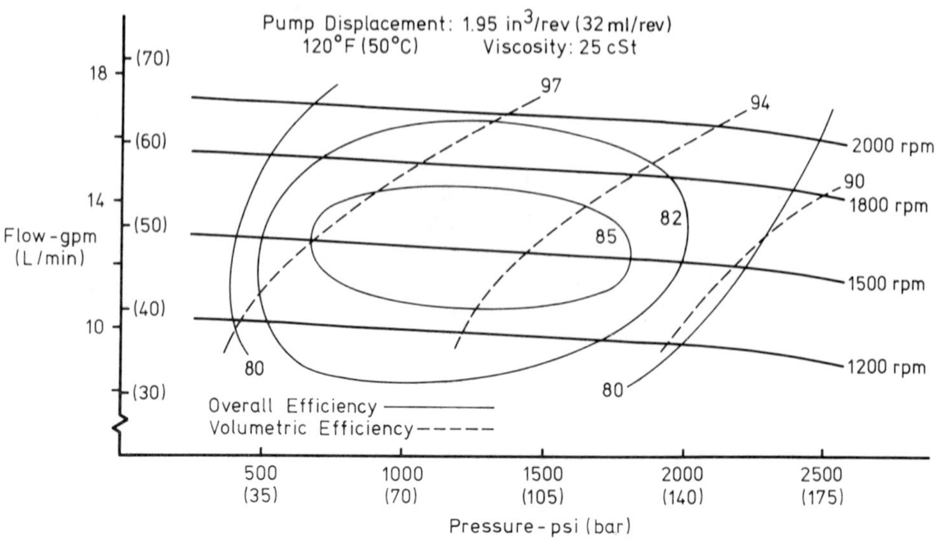

FIGURE 3-19 Single Eccentric Variable-Vane Pump Efficiency for Full-Flow Operation

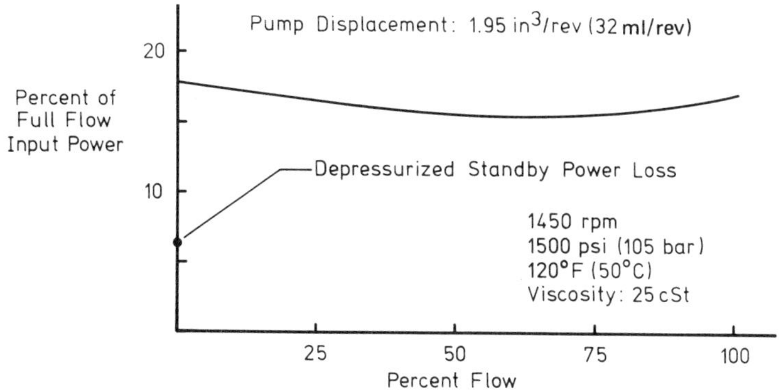

FIGURE 3-20 Single Eccentric Variable-Vane Pump Power Loss for Part-Flow Operation

Power loss tends to remain relatively constant over the flow range from full to zero flow. A typical curve is shown in Figure 3-20. If the pump is to remain in a zero flow, standby condition for any appreciable portion of its operating time, provisions for depressurizing the pump are desirable. As the figure shows, the depressurized loss at zero flow is about one-third that found at full pressure. This relationship holds reasonably well throughout the speed and pressure range of typical industrial applications. Means for depressurizing through the use of dual range controls or load-sensing pressure compensators is discussed in the next chapter.

The power loss within a pump shows up as heat added to the oil. Most of this heat is carried away in the outlet flow of the variable vane-pump when the flow is in the middle to upper part of the displacement range. There is also a cooling flowport, comparable to the case drainport of piston units. This cooling flowport ensures sufficient circulation through the pump and back to reservoir in the lower part of the displacement range down to zero flow to keep the pump from overheating.

C. Application Characteristics

Single eccentric variable-vane pumps have several characteristics that make them attractive for industrial applications.

Their sound level is low. The pump of Figure 3-19 and Figure 3-20 has a sound level of 65 dB(A) over the range of flows from full to zero at 2500 psi (175 bar) and 1450 rpm. At lower pressures, the sound level is correspondingly lower.

They offer a power saving, in many systems, as compared to fixed displacement units.

They are available with a wide range of control options and can be adapted readily to the diverse requirements of industrial control systems. In addition to such control arrangements as pressure limiting, load-sensing and displacement controls, this includes also the miscellaneous control configurations described in Section IX of Chapter 4.

Since they are short in length and have a through-drive shaft, they are readily adaptable to stacking in multiple configurations and driven off a single-drive pad.

The inlet-fill characteristics of this type of pump are good and they will operate at atmospheric inlet pressure or slightly below, with no cavitation through the rated speed range.

V. BALANCED VARIABLE VANE PUMP

A. Operating Principle

The balanced vane pump described in Section III of Chapter 2 is also available as a variable-displacement unit. In its variable configuration, it retains the feature of completely balanced, radial-pressure forces on the rotor so that the shaft bearings are completely unloaded. This permits reliable operation at higher pressures than the unbalanced single eccentric design allows. The fact that there are two pumping actions per revolution also tends to make it more compact.

The balanced variable-vane pump retains the low sound level inherent in vane designs and is therefore particularly suited to any industrial application. It is also adaptable to through drive construction for the creation of multiple pump packages.

The balanced variable-vane design has a single rotor, but there are two vanes in each rotor slot, side by side. Each vane runs on its own cam ring. The rotor and vanes are shown in Figure 3-21 and the cam rings in Figure 3-22.

When the two cams are aligned, as shown in the upper part of Figure 3-23, the two vanes in each slot work together. All vanes are at maximum extension over the pumping region and at minimum extension over the sealing region (the shaded areas of the figure). The difference in extension of the pumping and sealing vanes as they rotate produces flow.

As the cams are rotated, the average extension of the pumping vane is reduced and the average extension of the sealing vane is increased. When rotation of the cam surface reaches 90°, the average extensions are equal and no net flow is produced.

FIGURE 3-21 Balanced Variable-Vane Pump Rotor and Vanes

FIGURE 3-22 Balanced Variable-Vane Pump Cam Rings

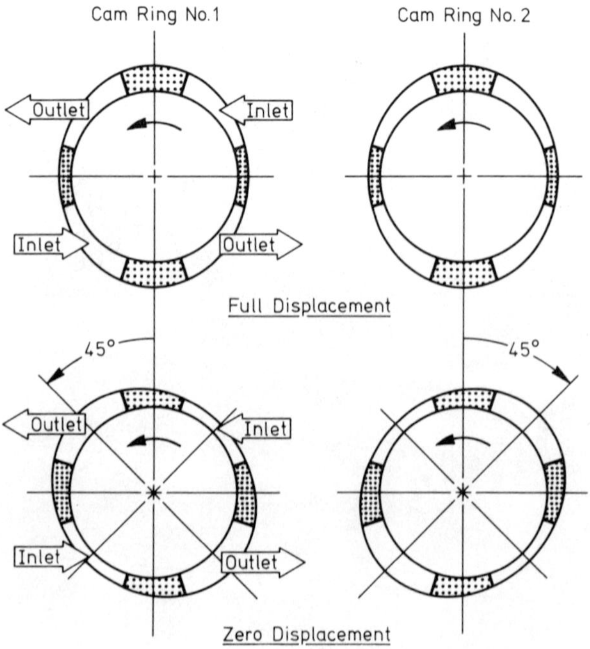

FIGURE 3-23 Balanced Variable-Vane Pump Operating Scheme

In the actual design, the cams are rotated in opposite directions in going from full flow toward zero flow, as indicated in the lower part of Figure 3-23. Thus, as the vanes pass through a pumping or sealing area in part-flow operation, the outward motion of one is balanced by the inward motion of the other. This action keeps the volume of fluid between successive pairs of vanes constant and produces a smooth, pulseless flow at all displacements.

To synchronize the cams and produce rotation, each cam has gear teeth on a segment of its outside diameter. These mesh with gears in the housing, as shown in Figure 3-24, which in turn, are rotated in opposite directions by a hydraulic actuator which moves a rack. The gears are spring-loaded in the full-flow direction and are also urged toward full-flow by a bias piston.

The control of pump displacement is normally through a pressure compensator which acts exactly as the one described in Section IIB for piston pumps. An overall view of the complete pump is shown in Figure 3-25.

B. Performance

At industrial operating conditions, the balanced variable-vane pump efficiency is similar to that found in piston units. The variable-vane pump has a pressure-

FIGURE 3-24 Balanced Variable-Vane Pump Cam-Rotating Mechanism

loaded flex plate construction of the type used in the fixed-displacement, balanced-vane pump described in Chapter 2. This controls the running clearances and results in high volumetric efficiency, generally above 97%. Overall efficiency at full-flow approaches 90% over the upper end of the operating pressure range.

Figure 3-26 is a typical efficiency map of a balanced variable-vane pump. At an operating speed of 1200 rpm, which is typical of a large percentage of industrial applications, overall efficiency is 89% or above over the range of pressures from 1500 psi to nearly 3000 psi (105 to 210 bar).

Part-flow performance is shown in Figure 3-27 for a typical operating condition of 1200 rpm and 2500 psi (175 bar). Power loss is less than 10% of the

FIGURE 3-25 Balanced Variable-Vane Pump Assembly

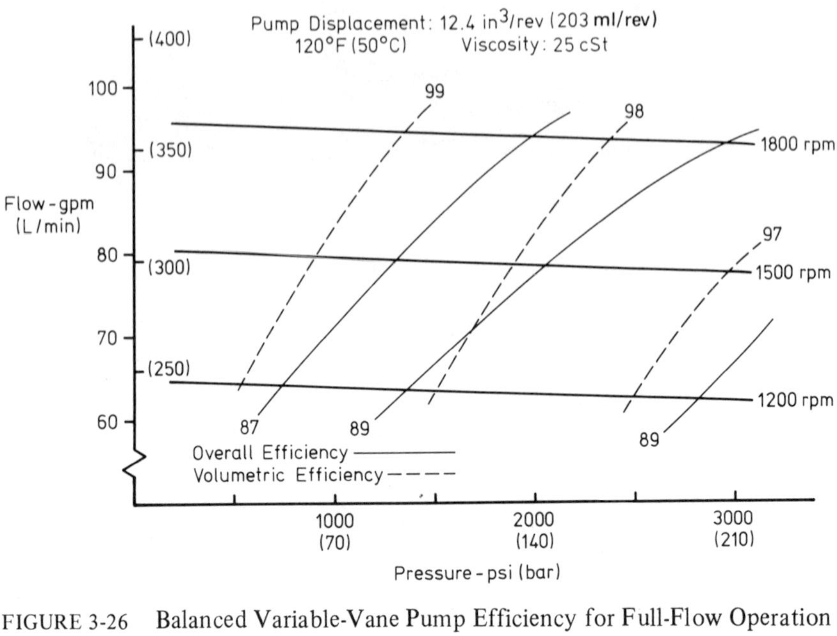

FIGURE 3-26 Balanced Variable-Vane Pump Efficiency for Full-Flow Operation

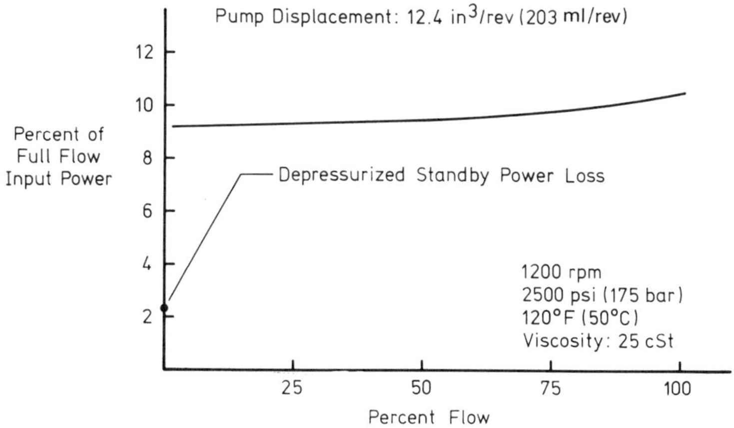

FIGURE 3-27 Balanced Variable-Vane Pump Power Loss for Part-Flow Operation

full-flow input power over most of the flow range. Standby power loss is also very low—near 2%—when the pump is operated in a zero-flow, depressurized mode. This is accomplished by using a load-sensing pressure compensator or one of the other depressurization schemes described in Chapter 4.

C. Application Characteristics

The overall design of the balanced variable-vane pump involves the use of a through shaft. The pump can therefore be adapted to through-drive operation with a second unit mounted on it and driven by the variable pump shaft. Drive-shaft torque capacity is adequate to drive a second unit (or combination of units) with hydraulic power capability equal to the variable pump.

Through-drive capability permits a variety of variable/variable or variable/fixed pump combinations to suit various system needs. This includes the addition of a small-fixed or variable pump as a source of auxiliary pilot pressure for servo systems.

As with all other variable pumps, it is necessary to carry cooling flow from the pump back to the reservoir under low- or zero-flow conditions. A cooling flowport is provided for this purpose.

The flow of the balanced variable-vane pump can be controlled by a pressure compensator or by any of the other variable pump control schemes described in Chapter 4. This includes the electrohydraulic control of displacement in electronically controlled systems.

4
Pump Controls and Systems

I. INTRODUCTION

For many years, the innovations in hydraulic power control were led by the aircraft industry. As aircraft became larger and flew faster, it became essential to amplify the muscle power of the pilot to overcome the high aerodynamic forces on the control surfaces. While there has been a continuing challenge to the aircraft hydraulic-component designer to make the elements of the system smaller, lighter, more responsive and more efficient, the systems themselves quickly gravitated to a central hydraulic power supply with individual control surfaces or other functions served from a constantly pressurized manifold. This, incidentally, is the type of system now commonly used in hydraulically powered robots: A central source of hydraulic power with individual servos for the multiple axes.

In actual aircraft systems, there are multiple circuits of this kind with multiple pumping sources offering a high degree of redundancy and complete safety from even multiple modes of failure. Nevertheless, each individual circuit is of relatively simple form and the pump and motor controls, while greatly refined, do not offer a great deal of variety in approach.

In more recent years, the role of leadership in hydraulic power application has been taken over primarily by the designers of some of the large earthmoving and construction equipment. The total power requirements of these machines are often well beyond those seen in aircraft. In addition, the concentration of power needs in only a few large loads, rather than a multitude of small ones, offers opportunities for new creative ways to distribute and control hydraulic power. Similar demands (and opportunities) have also occurred here and there in industrial hydraulics.

The innovative controls worked out for these new types of applications are now providing new levels of productivity, controllability and economy in many machines. Their application is limited only by the ingenuity of the system designer. This chapter will present a discussion of the controls now commonly available and some guidance in how they may be best applied.

II. SYSTEM EFFICIENCY

In earlier chapters, we discussed pump efficiency. In selecting and applying pumps in actual systems, power losses which occur within the system can completely overshadow those within the pump itself. Selection of the type of system to be used (and the type of control for variable pumps) can, therefore, be of overriding importance if the system is to be energy-efficient as well as responsive and reliable.

In hydraulics, the power delivered is proportional to pressure (p) times flow (q_V). If there is a pressure drop in the system or due to the throttling of flow through a valve, the energy loss is determined by $\Delta p \times q_V$. The wasted energy appears as heat in the fluid. If the flow is forced across a relief valve, the total output power of the pump is lost and heating can become very rapid.

Leakage within the system also results in lost power and in heat generation. The losses in this case are proportional to $p \times q_V$ (leakage). The objective of good system design, therefore, in addition to providing the necessary functional control of the pump's hydraulic power output, is to avoid unnecessary pressure drops and pressure-flow losses.

III. TYPES OF HYDRAULIC CIRCUITS

Before looking at the alternatives available in system design and pump selection, it is first necessary to define some of the terms used to identify basic types of directional control valves and systems. The directional valve which is used to control the flow from the pump may be of either open-center or closed-center design.

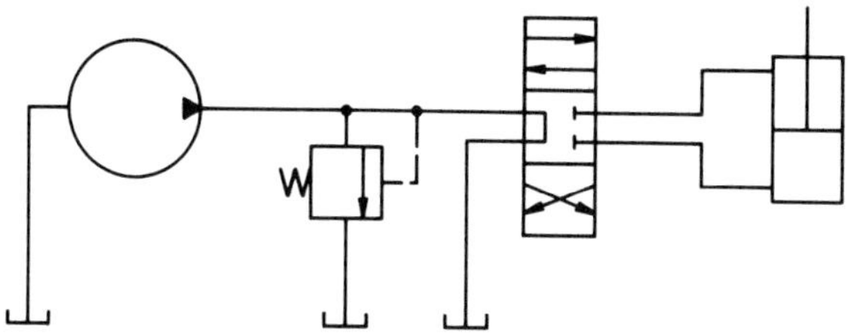

FIGURE 4-1 Constant Flow System with Open-Center Directional Valve

A. Open-Center Directional Valve

The open-center valve is illustrated in Figure 4-1. When it is in the centered position, it returns the output of the pump directly to the reservoir with minimum energy loss. Shifting the valve in either direction controls the motion of the load. This type of system is commonly used with fixed displacement pumps and occasionally with variable pumps. The load may be a cylinder, a motor or any other hydraulic device. This type of system is said to be *load-sensitive*. That is, the pump delivers only the pressure required to move the load, plus the pressure drop across the directional valve, of course. When the valve is fully shifted and calling for maximum flow, this pressure drop is relatively small and little energy is lost.

B. Closed-Center Directional Valve

A closed-center valve is illustrated in Figure 4-2. When it is in the centered position, the flow from the pump is shut off. With a fixed displacement pump, this would force the entire flow across the system relief valve which would be an obvious major waste of energy. Variable pumps, on the other hand, are commonly equipped with a pressure-limiting control or "pressure compensator." The

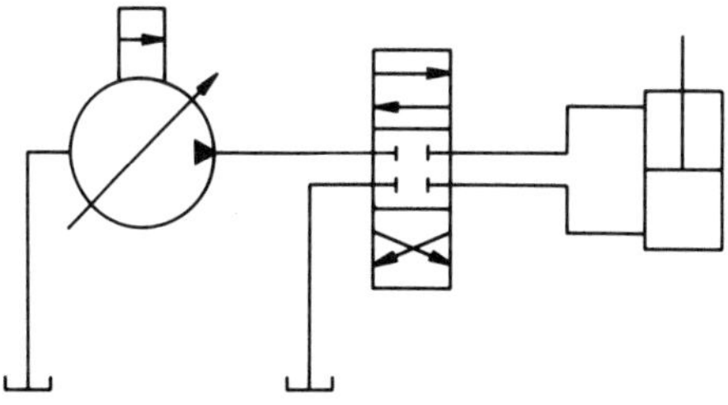

FIGURE 4-2 Constant Pressure System with Closed-Center Directional Valve

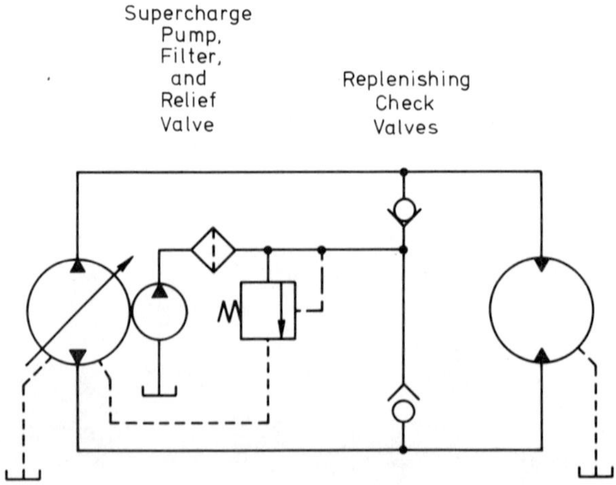

FIGURE 4-3 Closed-Loop System

control senses pump outlet pressure and automatically moves the pump to the zero-flow position when the predetermined setting of the control is reached. Shifting the directional valve in either direction causes the pump outlet pressure to drop. The compensator then automatically increases pump displacement, causing the load to move in the indicated direction.

In the system of Figure 4-1, if the load is stalled while the directional valve is shifted, such as when an actuator cylinder bottoms out, the system pressure reaches the relief-valve setting. High power loss results until the directional valve is recentered. In the system of Figure 4-2, with a pressure-compensated variable pump, if the load stalls, the pump will return to zero flow, minimizing system power loss.

C. Open-Loop and Closed-Loop Systems

The systems pictured in Figures 4-1 and 4-2 are known as open-loop systems. The pump receives fluid from a reservoir and the fluid returning from the load or by-passed by the open-center directional valve is returned to the reservoir.

In contrast, a closed-loop system is one in which the oil returning from the load goes directly to the pump inlet. The closed hydraulic loop is normally kept filled by a separate supercharge pump. A typical system is shown in Figure 4-3. The load is generally a motor capable of continuous rotation.

The pump is normally a variable displacement unit, to permit variation in the speed of the motor, and is usually capable of going "over center" to change the direction of flow and reverse the direction of motor rotation. This type of closed-loop system will be discussed more fully in the chapter covering transmissions.

IV. POWER LOSS DURING ON-OFF OPERATION

Many hydraulic systems have a simple "on-off" function. The load either is being moved at maximum velocity or is stationary with the pump unloaded. For this type of operation, the simple open-loop, open-center system of Figure 4-1 is quite adequate and very efficient. A variation of this duty cycle that occurs frequently involves moving a load at full velocity followed by a holding operation with no motion of the load. In the system of Figure 4-1, the full pump output would then pass across the relief valve, causing an extreme waste of power.

One type of system that reduces this high-power loss is shown in Figure 4-4. A dual pump in which the larger unit is unloaded during the holding operation will avoid most of the power loss. The unloading can be accomplished with an automatic, pressure-operated unloading valve. An alternate would be to use a large variable-displacement unit which automatically goes to zero flow during holding. This was previously shown in Figure 4-2. Large variable units can be expensive, however, and standby losses at zero flow and full pressure can exceed those in an unloaded, fixed displacement pump. The most cost-effective and energy-effective solution then may well be the use of a large fixed unit together with a small variable as shown in Figure 4-5. The large unit would again be unloaded automatically when holding.

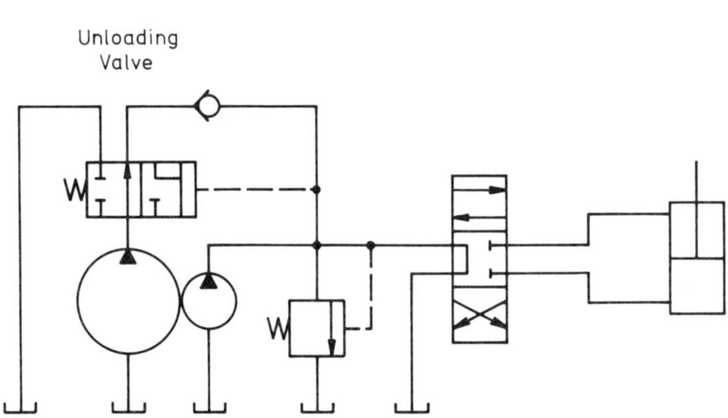

FIGURE 4-4 Fixed-Pump On–Off System with Unloading Valve

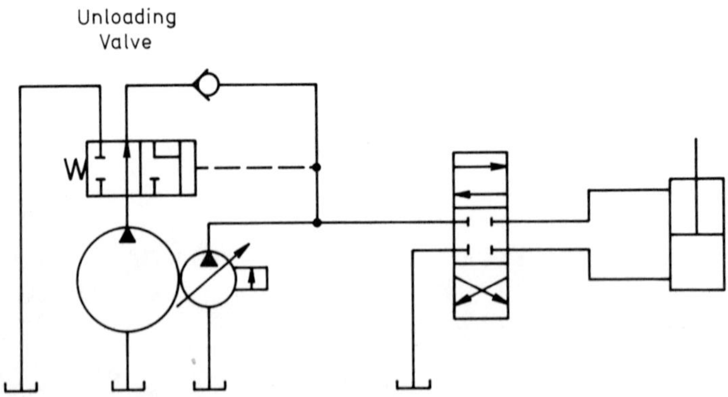

FIGURE 4-5 Fixed- and Variable-Pump On-Off System with Unloading Valve

V. POWER LOSS DURING PARTIAL FLOWS

A more complicated type of problem occurs when it is desired to control the speed of motion of the load over a wide range. The problem is further complicated when the magnitude of the load is also variable. To understand the consequences of the various choices, it is appropriate at this point to introduce the concept of a pressure/flow diagram.

A. Pressure/Flow Diagram

A typical pressure/flow diagram is shown in Figure 4-6. In the diagram, the maximum flow is established by the displacement and operating speed of the pump. The maximum pressure is determined by either the system relief valve or the variable pump compensator setting.

Individual load demands may fall anywhere within this pressure/flow diagram. Remembering that power is proportional to pressure times flow, the maximum possible power output is represented by point C (called "corner power") and the magnitude of the maximum power is directly proportional to the full area bounded by the limiting flow and pressure lines. If point L represents the pressure required to move a load and the flow needed to impart the desired velocity, the actual power required to move the load is, therefore, only the shaded area—far less than the total available.

B. System Power Loss with Open-Center Valve

When a system of the type shown in Figure 4-1 is operated at partial flow such as point L in Figure 4-6, the pump itself will be operating at point A. It will deliver full flow to the directional valve at the pressure required by the load. If

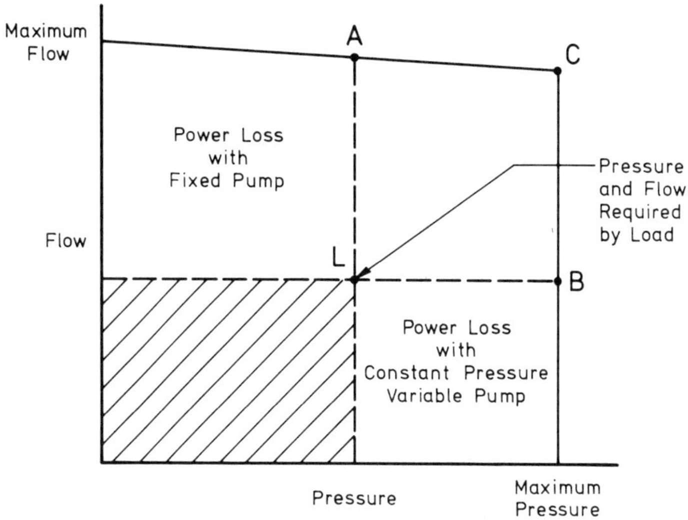

FIGURE 4-6 Pressure/Flow Diagram

only part of the flow is metered to the load, the remainder is bypassed to the reservoir at full-load pressure, resulting in the system power loss indicated in Figure 4-6. This is, of course, a loss chargeable to the system and it is independent of and in addition to any losses within the pump itself.

Pressure-compensated variable pumps are occasionally used with open-center directional valves. Unless the pump has some other type of special control (such as torque-limiting), there is no real advantage gained in a partial flow condition. Until the compensator setting (point C in Figure 4-6) is reached, the pump will remain at full flow (point A) and the system power loss will be the same as with a fixed pump. There is, of course, a considerable saving if the load bottoms out while the directional valve is in the shifted position. A fixed displacement pump will pass all of its flow over the system relief valve at full-system pressure while the variable pump will merely go to zero flow.

C. System Power Loss with Constant-Pressure Variable Pump and Closed-Center Valve

In a constant-pressure system of the type shown in Figure 4-2, the pump will remain at the maximum pressure level of Figure 4-6 regardless of the load demand. It will adjust its flow when the directional valve is in a partial flow-metering position and will operate at point B in the diagram. When the pressure required by the load is below maximum, however, the difference between pump pressure and load pressure appears as a pressure drop across the directional valve.

This pressure drop times the flow delivered to the load, is the power loss represented by the area at the right hand side of Figure 4-6. When flow is high and pressure is low, it can be a sizable loss.

D. System Power Loss with Load-Sensing Variable Displacement Pump and Closed-Center Valve

The potentially-large power losses associated with partial flow operation in either of the above systems can be avoided by the use of a load-sensing variable displacement pump control. When applied as shown in Figure 4-7, the actual pressure required to move the load is fed back to the pump stroke control. The pump then automatically maintains an output pressure that is usually in the range of 100 to 300 psi (7 to 20 bar) higher than the load pressure. This differential (which is adjustable up or down) is sufficient to maintain flow through the directional valve to the load.

A pressure/flow diagram for a partial flow condition is shown in Figure 4-8. The variable displacement pump adjusts its flow to that called for by the directional valve. It also adjusts its pressure output to the preset differential above the load pressure. The product of the pump flow and this differential pressure results in the power loss indicated in Figure 4-8—much smaller than in either of the previous systems.

The load-sensing pump will automatically adjust itself to any load demand within the flow and pressure boundaries of the pressure/flow diagram. The load-sensing pump control is usually used in combination with a pressure-limiting stroke control, which will act as an override to reduce stroke when the system maximum pressure is reached.

VI. TOTAL SYSTEM EFFICIENCY

While the constant-flow and constant-pressure systems of Figure 4-1 and Figure 4-2 have potentially high power loss in certain operating conditions, there are,

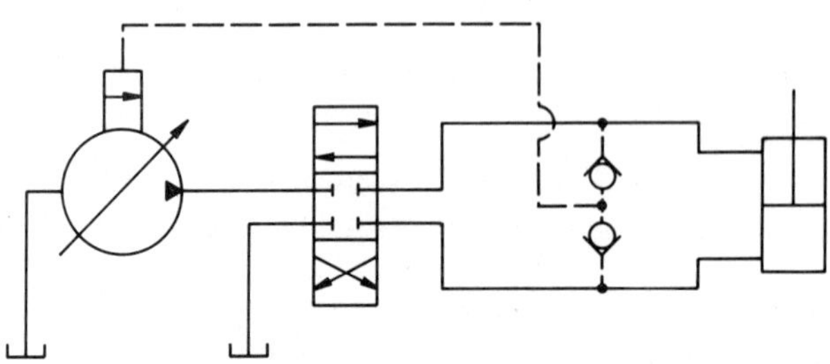

FIGURE 4-7 Variable Displacement Pump with Load-Sensing Control

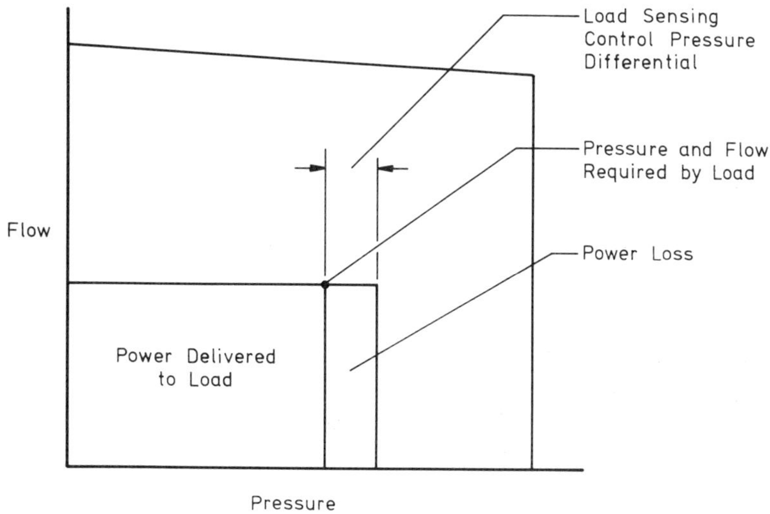

FIGURE 4-8 Pressure/Flow Diagram for Load-Sensing Pump

nevertheless, broad regimes of operation where the system efficiencies are quite acceptable. Before selecting either of those systems, or the load-sensing system of Figure 4-7, it is desirable to determine the characteristics of the load to be handled, such as:

Is the load usually light, with only occasional maximum requirements, and is it usually moved at maximum velocity?

or, in contrast,

Is the load often moved very slowly at well below maximum speed, and does it frequently require pressures near the maximum limit?

or, perhaps,

Is the load completely random, having no particular pattern, and requiring all combinations of pressure and flow?

The influence of these and similar factors on system selection can be assessed by means of the following total system efficiency plots. They include not only the influence of system losses as shown in Figures 4-6 and 4-8, but also the internal pump losses. The plots are based on actual data for mobile systems operating at 3000 psi (210 bar) and around 100 hp (75 kW). The patterns of efficiency are so distinct, however, that they can be considered broadly applicable to virtually any mobile or industrial system, regardless of pressure or power level or the efficiencies of the components that are chosen.

A. Constant-Flow System with Open-Center Valve

The pattern of efficiency for partial flow operation of a constant flow system is shown in Figure 4-9. The efficiency is very good at high flows, is reasonable at flows above 60% of maximum flow, and is only slightly affected by pressure. The system is simple and is reasonably efficient when load velocities are characteristically high and when there is to be little low speed maneuvering.

B. Constant-Pressure System with Closed-Center Valve

In contrast to the above system, the constant-pressure system efficiencies displayed in Figure 4-10 show an almost opposite pattern. While efficiency is high at high flows, as it is with a constant flow system, it falls off rapidly at lower flows as pressure is reduced. This system is at its best when loads are characteristically high, and low-speed operation is frequent, although it also does well at nearly any very high-flow condition.

C. Load-Sensing System with Closed-Center Valve

The load-sensing system, because of its flexibility in adapting to any variation in flow or pressure demands, has a distinctly different pattern from the previous

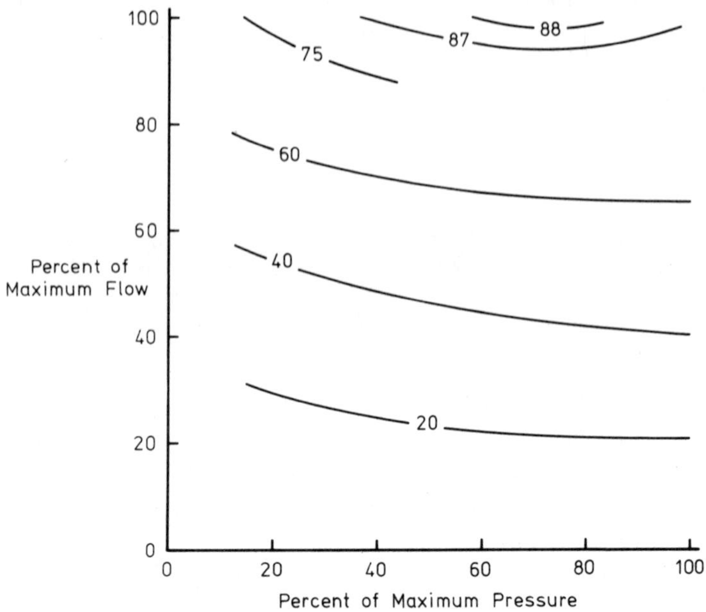

FIGURE 4-9 Constant-Flow System Efficiency Pattern

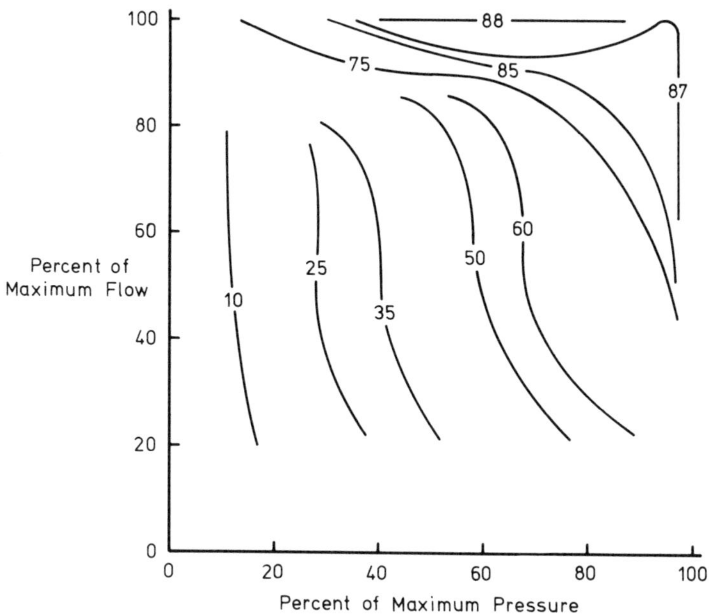

FIGURE 4-10 Constant-Pressure System Efficiency Pattern

two as is seen in Figure 4-11. While there is some reduction in efficiency as the overall power level drops, due either to flow or pressure reduction, the performance of the system is remarkably uniform over a broad range.

At 80% of maximum pressure and 20% of maximum flow, it will operate at about 70% efficiency as compared to about 20% for a constant-flow system or 50% for constant pressure. Near the other corner of the chart at 20% pressure and 80% flow, efficiency is near 80% as compared to 65% and 20% for the others. At a low-middle position of 30% of both flow and pressure, load-sensing system efficiency is still 75% compared to around 25% for the other two.

The load-sensing system is thus seen to offer a major energy saving in systems that are subject to wide fluctuations in load demand, either in flow or pressure. When coupled with a high-average power demand, the reduction in energy costs can be very significant.

D. Standby Power Loss

One last factor that must be considered in system power consumption is the level of standby-power loss. Most systems have periods of idle time where the pump continues to be driven by the prime mover but its power is not being used. If these periods are long, the total power consumption can exceed that used during a working cycle.

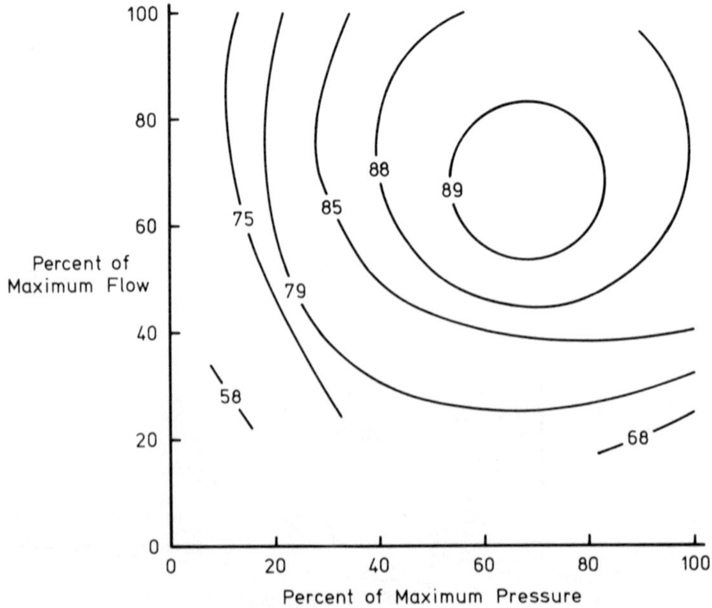

FIGURE 4-11 Load-Sensing System Efficiency Pattern

In fixed displacement systems with open center valves, the standby loss is influenced by the characteristics of the directional valve and by the length and size of the return line to the reservoir. The pressure drops through the valve and lines are each likely to account for some one or two percent of the full power capability of the system. Coupled with a probable 5% level of unloaded rotational losses in a vane or gear pump, a total of some 7% to 10% is likely for the entire system. A system having a full flow/full pressure (corner power) level of 100 hp (75 kW) is thus likely to have standby losses of around 7 to 10 hp (5 to 7.5 kW).

The constant-pressure system using a closed-center valve avoids any valve- or line-pressure drops during zero-flow standby operation. The pump itself remains in a pressure-loaded condition, however, and a pressure compensated piston pump will usually be found to have standby losses in the same 7% to 10% range as the fixed-flow, open-center system.

The load-sensing system is far superior. Not only are the valve and line losses eliminated as in the constant-pressure system, but variable-displacement pump losses are significantly lowered as well. At around the 100 to 200 hp level (75 to 150 kW), standby power loss has been found to fall in the 2 to 3% range. This can lead to a large total power saving over the life of the unit.

The values quoted above will vary somewhat with individual unit designs, but can be considered typical of fixed-displacement piston units operating at 3000 psi (210 bar) or above. At lower pressures, standby flow and rotational losses are not greatly reduced and constitute a higher percentage of the system full flow/full pressure power. Internal pump losses may also be higher in other pump designs, but the relative advantage of the load-sensing system will be maintained.

VII. LOAD VELOCITY CONTROL

When operating in a partial-flow mode, it is often desirable to hold the load at relatively constant velocity. If the directional valve is held in a constant metering position and the load does not change, no problem is presented. If the load changes, however, neither the constant-flow nor the constant-pressure system will maintain constant load velocity without readjustment of the directional valve position.

With either of the above systems, if the load-pressure requirement increases, load velocity will be reduced for a constant valve position. In the open-center valve, the increased pressure causes more flow to be bypassed and less delivered to the load. With the closed-center valve and a constant-pressure pump, the increase in load pressure reduces the pressure drop across the directional valve and, therefore, also reduces the flow that is directed to the load.

The load-sensing system, in comparison, will maintain constant load velocity for any directional valve position. This will remain true for any load variation within the system pressure limit. The load-sensing pump maintains a constant pressure drop across the metering orifice of the directional valve, automatically adjusting its output pressure to compensate for changes in pressure at the load. The constant pressure drop across the valve maintains constant-flow and constant load velocity.

VIII. PRODUCTIVITY

Productivity is related to machine- or operation-cycle time, which in turn, is dependent on the velocity with which the load can be moved. It is often possible to select a pump, a pump control and a type of system that will maximize the work that can be performed with a given level of installed power. This can be especially important where loads are highly variable and where very high load demands occur intermittently throughout the operation.

Referring to the pressure/flow diagram in Figure 12 and remembering that flow times pressure represents power, a constant power line can be plotted

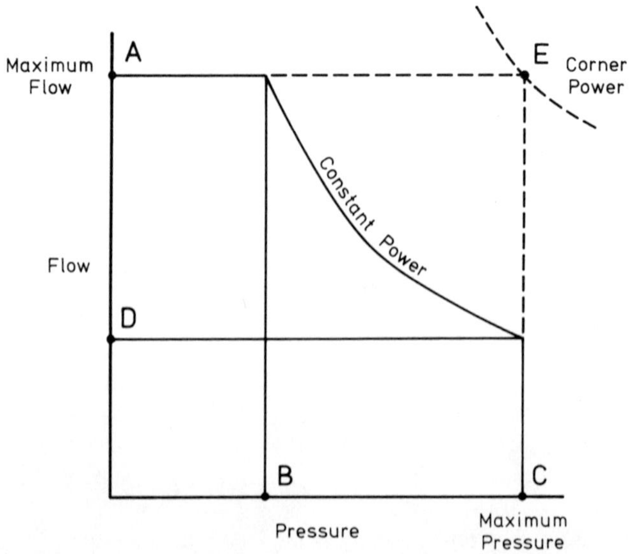

FIGURE 4-12 Pump Flows and Pressures for Constant Power

where $p \times q_v$ = constant. Strictly speaking, displacement per revolution times pressure represents pump torque and power is torque times rpm, but most systems operate at relatively constant speed, so that constant torque results in constant power. We will, accordingly, continue to use the two terms more or less interchangeably.

In Figure 4-12, if high flow is desired as at point *A*, the maximum pressure attainable within the power limit is represented by point *B*. Similarly, if the pressure represented by point *C* is needed on occasion to move a peak load, the maximum pump flow permissible is indicated by point D. To have the capability of providing both the maximum flow (A) and the maximum pressure (C), a single fixed pump, a pressure compensated pump or a load-sensing variable pump would require a corner power capability in the prime mover corresponding to point E—clearly beyond the desired power level.

A. Torque-Limiting Controls

The ability to operate at all points within the constant power curve is achieved by pumps with torque-limiting stroke controls. In such a control, the maximum stroke available in the pump is reduced as the pressure increases, preventing the pump from exceeding the desired maximum torque or power limit. For light loads however, high flows become available for rapid machine cycle times. This is possible while still retaining the essential ability to move much higher loads when necessary at a rate commensurate with the power available.

The maximum pressure in such a pump is usually established by a pressure compensator which overrides the constant torque control when the limiting pressure is reached.

Most variable displacement pump-torque limiting systems use closed-center valves identical with Figure 4-2. Occasionally however, the open center type of system illustrated in Figure 4-1 is used. There is no essential difference in the effectiveness of one system compared to the other when operating along the maximum power line with the valve fully shifted, as in Figure 4-13. Each will reach the same point on the maximum power curve and will deliver maximum power to the load, minus only valve and line losses.

There again is no practical difference during part-load operation when each type of system calls for pump operation on the maximum power curve. Figure 4-14 shows a typical example. In a closed-center system with the directional valve closed, the pump is at standby at full pressure at point A. As the valve is opened, the pump follows the maximum pressure and maximum power curves upward and to the left until the desired flow is reached at point B. The difference between the pressure at point B and the pressure required by the load results in a pressure drop across the directional valve and the indicated power loss.

With an open-center system, the pump is at standby at full flow and minimum pressure drop at point C. As the open-center directional valve is shifted,

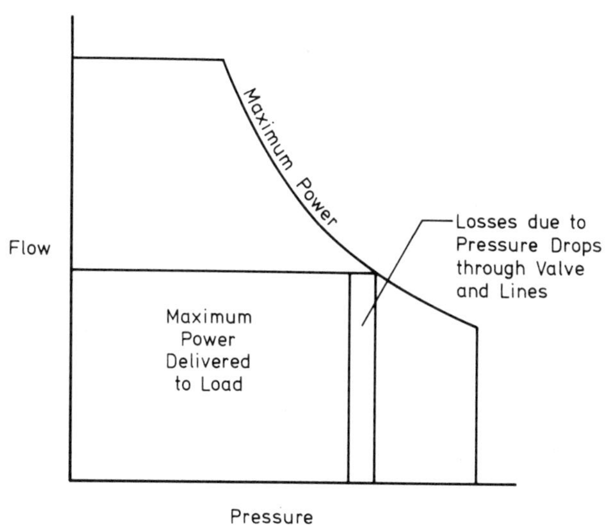

FIGURE 4-13 Torque-Limiting System Operating at Maximum Power Level

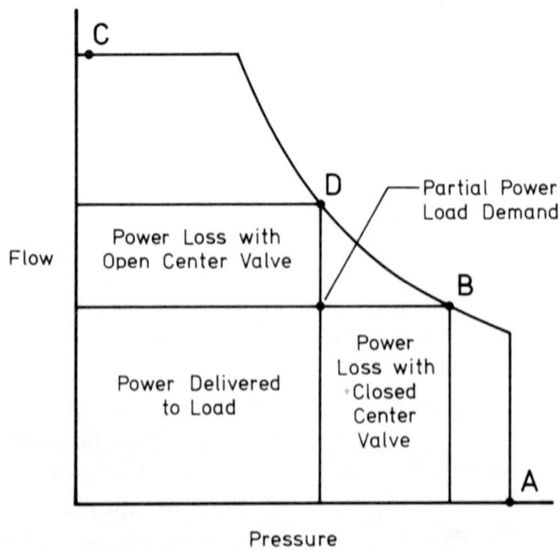

FIGURE 4-14 Torque-Limiting System at Partial Load Demand

pressure is directed to the load and the pump follows the maximum flow and maximum power lines to the right and downward to point D. The excess flow delivered by the pump is dumped at load pressure across the partially shifted open-center valve resulting in the indicated loss.

While the two rectangles representing losses in Figure 4-14 are of different shapes, the areas are identical and there is no difference in system power loss. This is easily seen if the load is considered to require 80% of maximum power. While each pump is at a different operating point on the power curve, each is delivering 100% of maximum power and the difference of 20% is lost.

A far different situation occurs, however, when it is necessary to handle high loads repeatedly at low speeds. In the closed-center system, as shown in Figure 4-15, the pump will be operating at its pressure compensated limit at point E and will be well below maximum power. While there will be a pressure drop across the directional valve when the load is less than maximum, the power lost will be small. The open-center system, in contrast, will have the pump operating at point F on the maximum power curve and the losses will be considerably larger.

B. Torque-Limiting System with Load Sensing

For maximum productivity and minimum energy consumption in a system that is to handle varying loads, a combination of torque-limiting and load-sensing is optimum. This entails adding a torque limiting control to the pump in Figure 4-7.

This combination of pump controls maintains the advantage of high productivity inherent in the torque-limiting system. It enables full use of power over the entire range of loads, giving maximum possible load velocity at all times. It also permits loads to be moved at less than maximum rate, when desired, with a minimum loss in power. Finally, it places the pump in a minimum power condition during standby.

The operating boundaries of a load-sensing/torque-limiting system are shown in Figure 4-16. The pump is still able to respond to full-load demands and operate at the boundaries of maximum flow, maximum pressure and maximum power. It can also leave those boundaries, however, when loads are low and a partial flow-metering position of the directional valve is employed. Figure 4-16 shows a partial load of the same magnitude as that of Figure 4-14. The power loss is clearly lower. Also, when the closed-center directional valve is returned to the neutral position, the pump goes to the zero flow-low pressure position represented by point G, yielding minimum standby losses.

C. Torque Summation

In some machines, the prime mover drives two large variable pumps. Frequently, the total power capability of the pump combination exceeds the power available from the prime mover. A typical example occurs in large construction machinery where all functions are hydraulically powered. While each pump can be provided

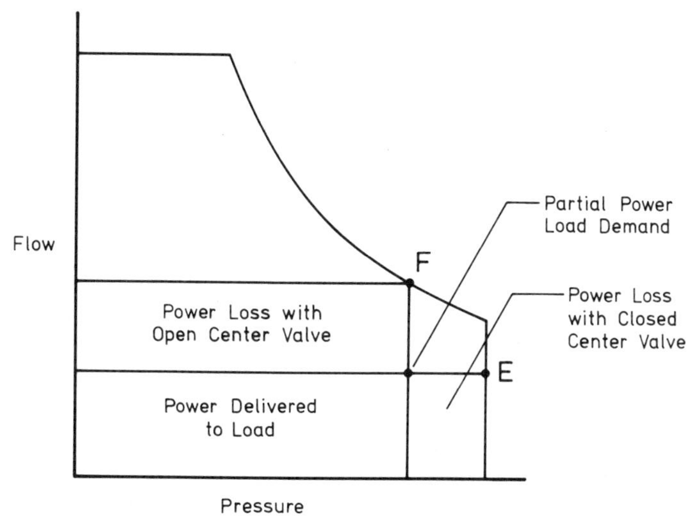

FIGURE 4-15 Torque-Limiting System During High Pressure, Low-Speed Operation

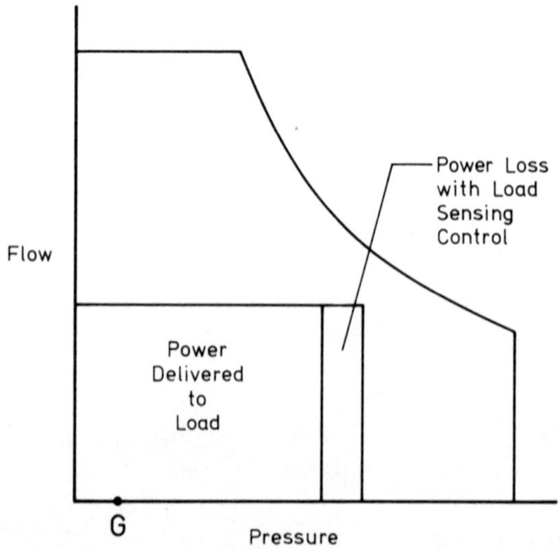

FIGURE 4-16 Torque-Limiting System with Load-Sensing Pump Control

with an individual torque-limiting control, this does not permit the optimum use of the available power.

As an example, if the prime mover has the capability of driving only one pump at its corner power, each pump's individual control would need to be set at the 50% power level to prevent the total from exceeding the prime mover's capability in the extreme case. This means that even if one pump is lightly loaded, the second will be limited to 50% power, and the remainder of the prime mover's power will not be available.

The full prime mover power can be made available and maximum productivity of the machine maintained when true torque summation is used. In true torque summation, each pump not only senses its own torque position (the combination of displacement and output pressure), but also that of the opposite pump. When one pump is operating at low power (below 50% in the above example), the second automatically raises its own torque limit to correspond. The pumps, as a pair, still operate within the power limit of the engine, but each is automatically free to make use of any increment of power that is not being used by the other.

Torque summation can be used together with load-sensing and pressure-limiting to retain the low power loss at part load conditions depicted in Figure 4-16. Torque summation supplements the load-sensing and pressure-limiting functions and does not replace them.

D. Speed Sensing

Some machines not only drive a multiple number of pumps but also draw power from the prime mover through some form of mechanical transmission. A typical example is in large wheeled loaders where the wheel drive is mechanical and all other machine functions are hydraulic. The maximum wheel load together with the installed hydraulic power capability can easily exceed the prime mover's maximum power output level.

Torque summation in this case is not appropriate since there is no ready way to obtain a torque signal from the transmission. However, in governed engines, the engine rpm begins to fall off from the governed speed when the engine power limit is approached. This fact permits the use of a speed-sensing control, which monitors engine speed and unloads the hydraulic pump when engine rpm begins to fall, allowing full power to be directed to the wheel drive for vehicle control.

An alternate use of speed sensing can occur in applications with large primary hydraulic power units that have a corner horsepower capability that exceeds the power limit of the prime mover. In this type of application, the speed-sensing control performs the same function as the torque limiting or torque summation controls described in Sections A and C. It limits pump stroke at each operating pressure level to stay within the power limit of the engine.

Speed-sensing pumps require a speed signal which is most readily obtained by monitoring the flow of a fixed displacement pump being directly driven by the engine. The flow of this pump can be directed past a moderate restriction to produce a pressure drop. The pressure drop in turn can be used as an actuating signal for an unloading valve in the pump's stroke control mechanism.

This type of control is needed most frequently in large, heavy-duty variable displacement piston pumps with high horsepower capability. This type of pump often has an integral, full-flow supercharge pump which offers a convenient source for the speed reference signal. The entire control, in fact, then becomes totally self-contained as part of the heavy duty pump package.

In systems with multiple variable pumps, the unloading can be applied to any one individually or to two or more, depending on system needs. When two or more pumps are to be unloaded, only a single fixed pump's flow is needed as a speed reference. Operation of either a single- or multiple-pump system is fully automatic, requiring no operator attention. It prevents engine stall and assures complete vehicle control.

The use of speed-sensing does not prevent the use of pressure-limiting, load-sensing, or torque limiting. It will override any of these however, when the engine's full power limit is approached.

E. Alternate Power-Limiting Systems

The advantages of a torque-limiting control are obvious in providing high rates of motion for high productivity with moderate loads while still maintaining high load capability for extreme cases. Variable displacement pumps of suitable size and ratings may not always be available however or may prove too expensive. It is therefore appropriate to consider other alternatives.

1. Wide Differential Compensator One alternative is a pump having a "wide differential compensator." This type of variable pump control is far simpler than a true torque-limiting control but produces a somewhat similar result. The pressure delivered by the pump at low flows can be considerably higher than that at full flow, giving some added load-carrying capability at low flows, as shown in Figure 4-17. This contrasts with the normal pressure-limiting compensator characteristic which provides nearly constant pressure between full flow and zero flow.

2. Combined Fixed/Variable Systems A further approximation of a torque-limiting variable pump control can be obtained by combining a fixed displacement pump with the wide differential variable unit. An automatic unloading valve, as shown in Figure 4-5, can be set to unload the fixed pump when the tor--torque limit of the prime mover is reached (point C in Figure 4-18). The variable pump is then available to provide its full flow to point A and partial flow to point B.

Combination fixed/variable packages with both pumps driven from a common input shaft are available in many cases. Figure 4-19 shows a fixed-displacement balanced vane pump with a variable piston pump mounted on it.

3. Staged Fixed Pumps For many applications, where the hydraulic power

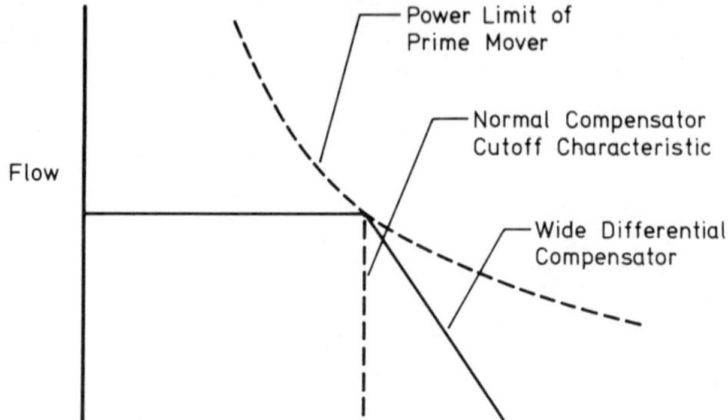

FIGURE 4-17 Pressure and Flow Limits for a Variable Pump with Wide Differential Compensator

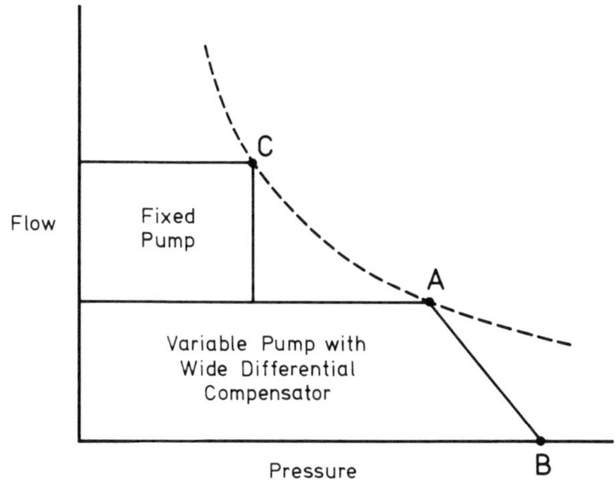

FIGURE 4-18 Combined Fixed/Variable System

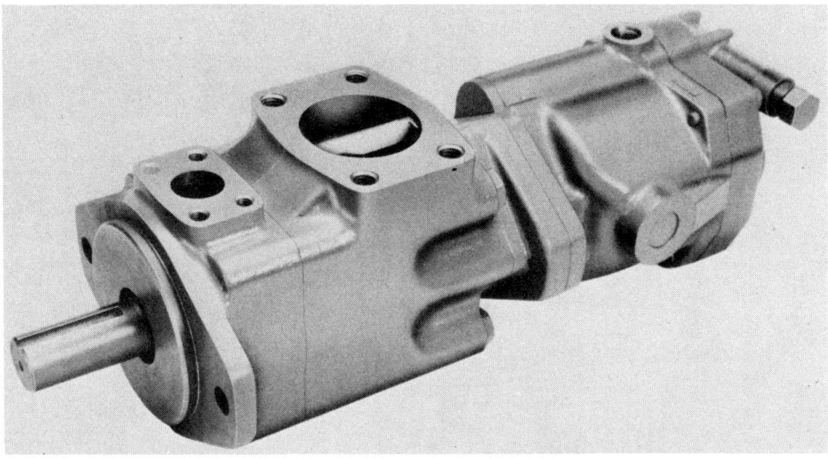

FIGURE 4-19 Fixed/Variable Pump Package

cycle involves a rapid traverse operation at low pressure followed by a high-pressure holding operation, a staged series of fixed pumps may be economical and energy-efficient. The characteristic pressure/flow chart of a triple pump combination, staged for maximum power availability is shown in Figure 4-20. A typical unloading valve arrangement is in Figure 4-21. The pumps in this case are selected to be of equal size.

Where the flow required for the holding operation is very small, the system power loss can be minimized by using a very small pump—either fixed or variable—to keep the system pressurized. The clamping pressure can also be

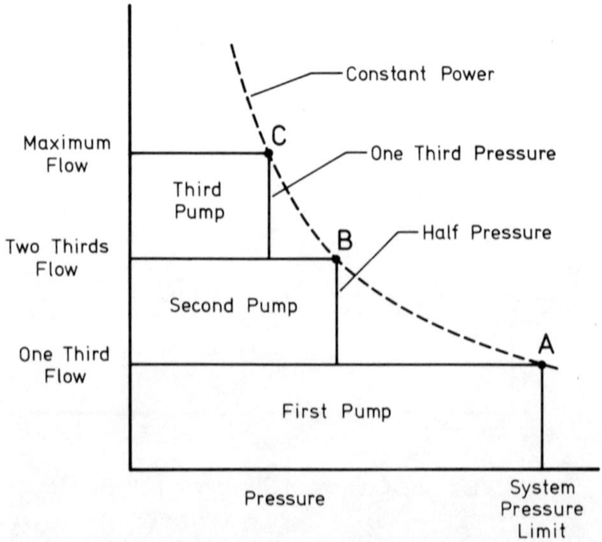

FIGURE 4-20 Staged Fixed Pumps with Constant Power Limit

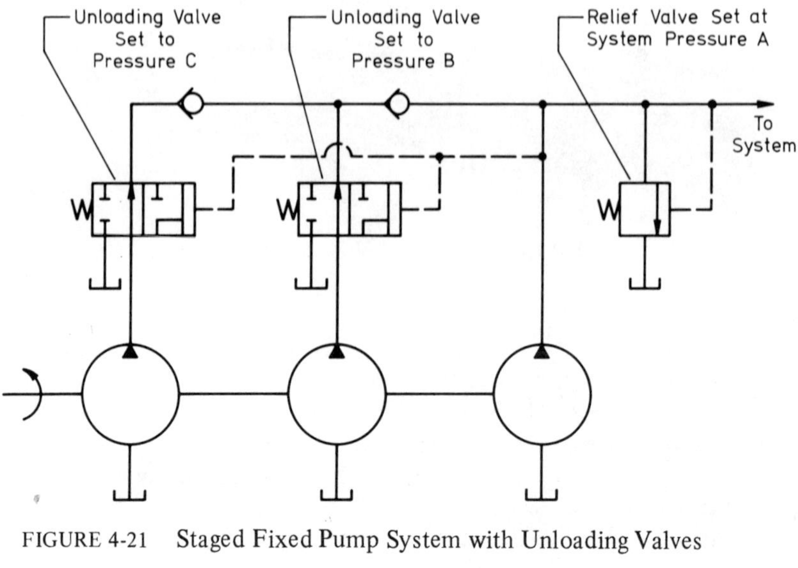

FIGURE 4-21 Staged Fixed Pump System with Unloading Valves

FIGURE 4-22 Triple-Balanced Vane Pumps with Common Inlet

increased accordingly, if needed, within the boundaries of the torque curve and within the pressure ratings of the pump and other system components.

Stacked groups of fixed displacement pumps, driven from a single input shaft, are available in groups of two, three, and sometimes more. Figure 4-22 shows a stack of three balanced vane pumps. This package is designed with a single-inlet connection to feed all three units, thus simplifying installation. Outlets are separate to permit either feeding individual loads or directing the entire output to a single load by way of unloading valves.

F. Displacement Controls

Some variable pumps are available with a control arrangement that permits the operator to select the displacement desired. The control may be manual, hydraulic, or electrical. When coupled to the load, either directly or through a completely-shifted directional valve, the system losses can be even lower than for a load-sensing system.

The directly coupled system is inherently load-sensitive, delivering only the pressure required by the load. The pump in turn delivers only the flow called for by the operator to achieve the required load velocity. The only system losses are line losses.

Systems that use displacement controls generally enjoy these advantages:

1. very low system losses;
2. low standby losses, since the pump can be depressurized;
3. smooth, stepless changes in load velocity;
4. nearly constant load velocity for a given displacement position;

regardless of load fluctuations—this is because of the high volumetric efficiency of most variable displacement pumps; and
5. adaptability to remote controls and to electronic control systems.

Possible disadvantages of displacement controls include:

1. relatively slow response to input signals in most servo-controlled systems;
2. the need for some auxiliary control to prevent operation at excessive system pressure if the load stalls—e.g., a compensator override is desirable to prevent continuous dumping of the pump flow across a system relief valve; and
3. possible need for an auxiliary pump to provide pilot pressure for servo-controls or to provide cooling flow through the variable pump if it is completely unloaded at standby.

Pumps with displacement controls find their most frequent use in transmissions and will be discussed more fully in that chapter.

IX. MISCELLANEOUS VARIABLE DISPLACEMENT PUMP CONTROLS

There are many variable pump controls with special characteristics that are frequently applied to satisfy special system needs. Many of them are aimed specifically at allowing the pump's pressure or flow characteristics to be altered by remote control—either hydraulically or electrically. This can be highly desirable in some machine operations to achieve accurate control and minimize power consumption.

In discussing these controls, it will be convenient to introduce the diagrammatic representation of a pressure compensator—the pressure sensing valve which automatically controls the pump-stroking mechanism to maintain a constant pressure output. Figure 4-23 shows a representation of a variable displacement pump in which the stroking mechanism is spring-loaded toward the full-flow position. The pressure compensator is shown spring-loaded toward the right, connecting the stroking actuator piston to drain and allowing the pump to remain in the full flow position. When pump output pressure reaches the spring setting of the compensator, the valve shifts to the left, admitting pump outlet pressure to the stroke control piston and causing the pump to move toward zero flow, holding outlet pressure at the desired maximum value. The pump's operating boundaries are on the familiar pressure/flow (p/q_v) diagram shown in Figure 4-23.

Using the same diagrammatic symbols, a load-sensing compensator arrangement is shown in Figure 4-24. Pump output pressure acts on the right end of the

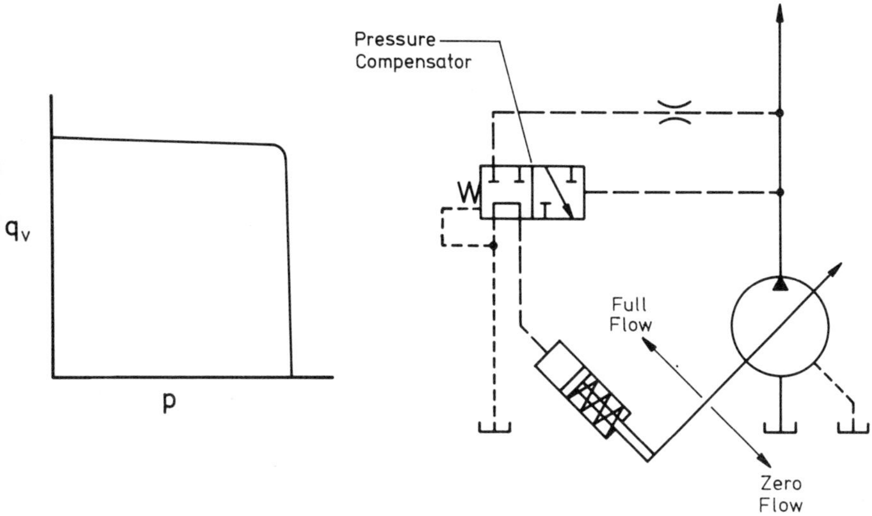

FIGURE 4-23 Functional Diagram of a Pressure-Limiting Compensator

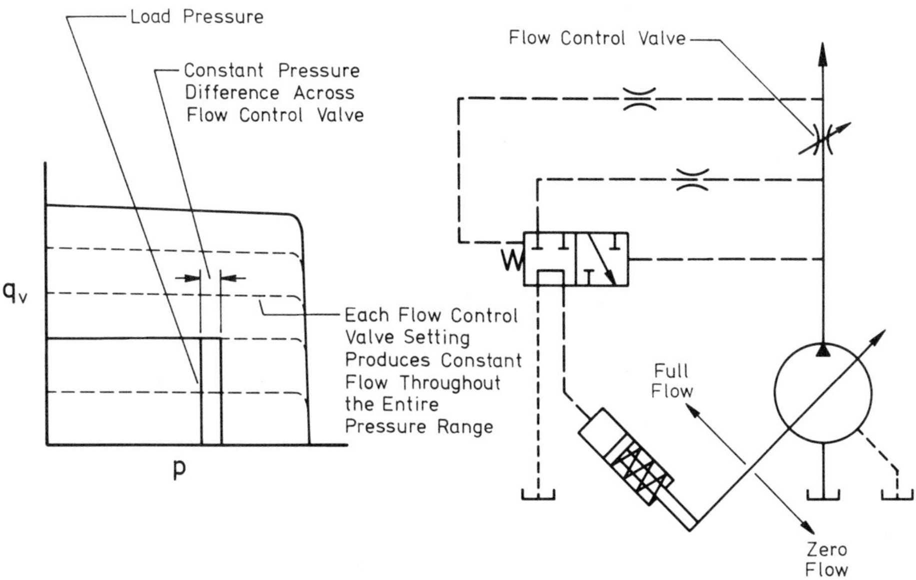

FIGURE 4-24 Functional Diagram of a Load-Sensing Compensator

compensator spool and load pressure is applied in the spring chamber. The compensator maintains pump outlet pressure at a fixed value above load pressure, as established by the compensator spring setting. For a given flow control valve setting, this constant pressure differential will maintain constant flow, regardless of load pressure variations.

These two simple controls of Figures 4-23 and 4-24 can be used to spawn a whole series of remotely controllable pressure and flow arrangements, all directly adaptable to electrically sequenced or computerized control systems.

A. Hydraulic Remote Control of Pump Pressure

In the diagram of Figure 4-23, the compensator spring chamber is connected to the drain. If, instead, it is connected to a remote source of pressure, as in Figure 4-25 (and a low value of spring setting is used), the pressure output of the pump can be varied over a broad range. At its minimum value of pressure, the pump can be unloaded for minimum standby power loss. Intermediate pressures up to maximum system pressure can be called upon for operating system needs.

B. Electrical Remote Control of Pump Pressure

The schematic diagram shown in Figure 4-25 requires a separate source of pilot pressure. There are also numerous alternate arrangements that are self-energizing, using the output pressure of the variable pump itself as the actuating means.

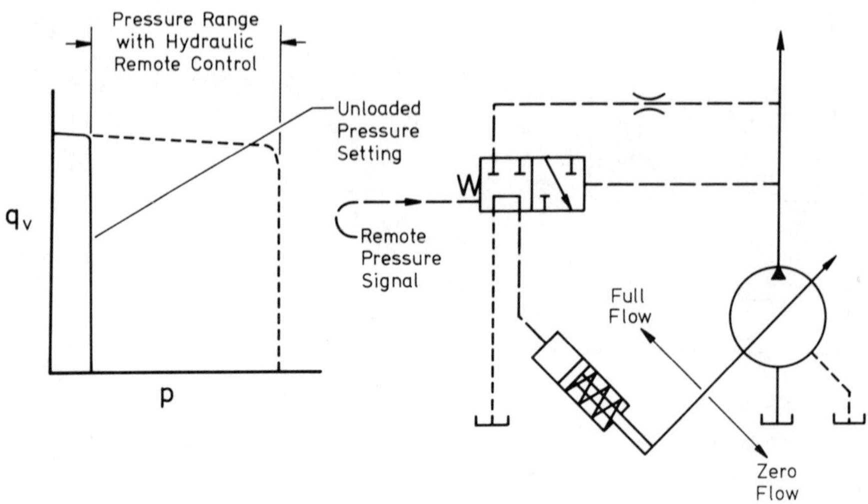

FIGURE 4-25 Hydraulic Remote Control of Pump Pressure

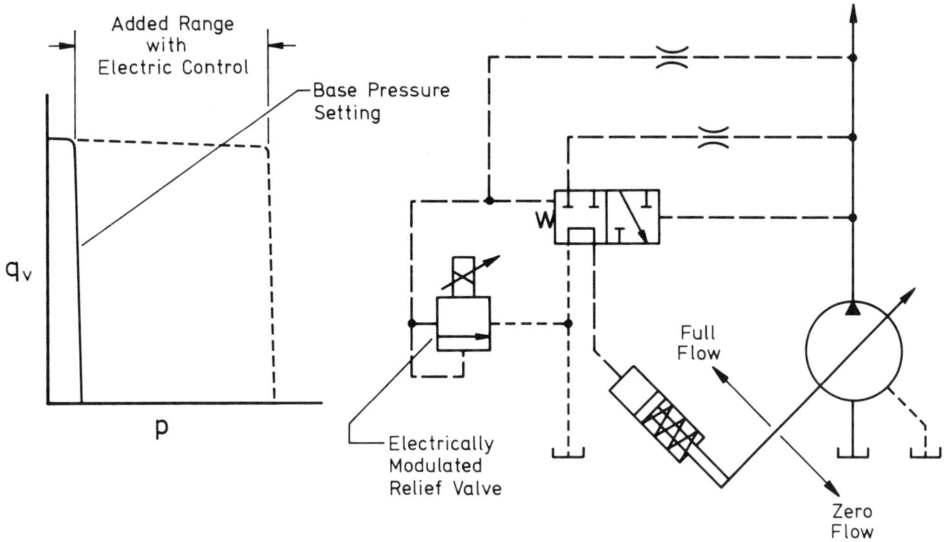

FIGURE 4-26 Electrical Remote Control of Pump Pressure

Figure 4-26 shows how pump pressure can be controlled from a remote location, using an electrically modulated relief valve. No auxiliary source of pilot hydraulic pressure is needed—the hydraulics is completely self-contained.

C. Dual Pressure Control

A means of remotely controlling pressure between two preselected levels is shown in Figure 4-27. The selector valve is electrically controlled and directs flow to either of two relief valves that establish the pressure in the pressure compensator spring chamber. If one of the two relief valves has a very low setting (or is replaced by a check valve), the pump's lower pressure level becomes an unloaded pressure.

D. Triple Pressure Control

A means for providing three discrete operating pressure levels is shown in Figure 4-28. Again, the actuation is electrical and no auxiliary pressure source is required. This arrangement is particularly useful where two system pressure levels are desired plus an unloaded standby pressure.

E. Electrically Modulated Flow Control

In the schematic diagram of Figure 4-24, each setting of the flow control valve will establish a constant value of flow to the system. The flow will be independent of system pressure, as indicated in the pressure/flow diagram at the left in the diagram. This type of controllability is often desired to permit a rapid change in system operating rates, as between a rapid traverse operation and a working feed rate. The insensitivity to system pressure, up to the maximum pressure limit, ensures uniformity of operation regardless of load level.

The flow control valve's metering orifice can be remotely controlled electrically or hydraulically, and makes the system readily adaptable to automatic control sequencing.

F. Constant Flow Control

In some applications, pump input rpm can vary over a broad range while system considerations require a constant pump flow. A typical example is the use of a variable displacement pump as a source of steering power on an internal combustion-driven vehicle. Pump rpm from idle to maximum engine speed can vary over about a four to one range. To provide adequate steering control at lower speeds without over control at high speeds, a constant flow to the power steering control valve is usually desired.

This is readily accomplished in the control arrangement of Figure 4-24 by using a constant orifice in place of the flow control valve. The orifice and all

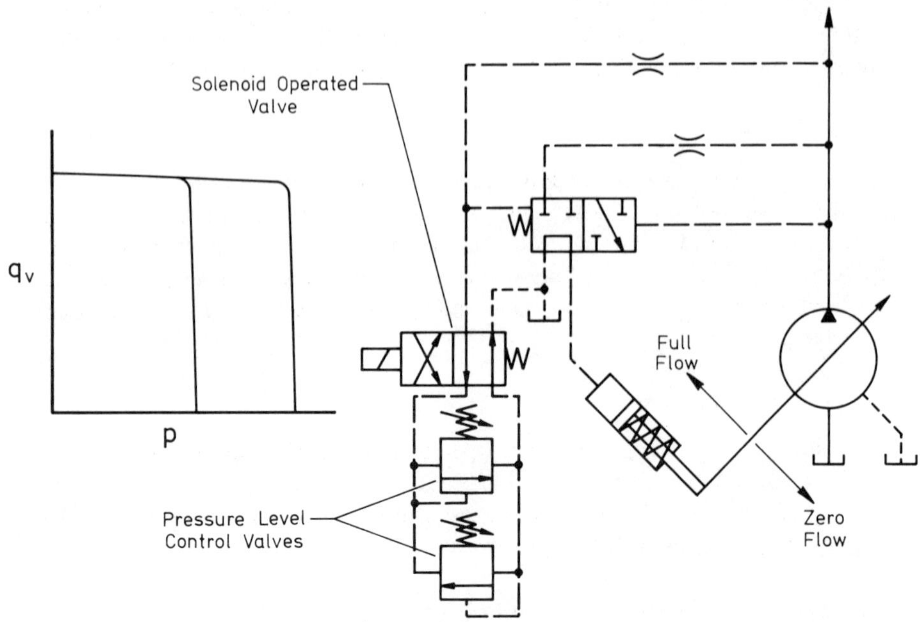

FIGURE 4-27 Dual Pressure Control

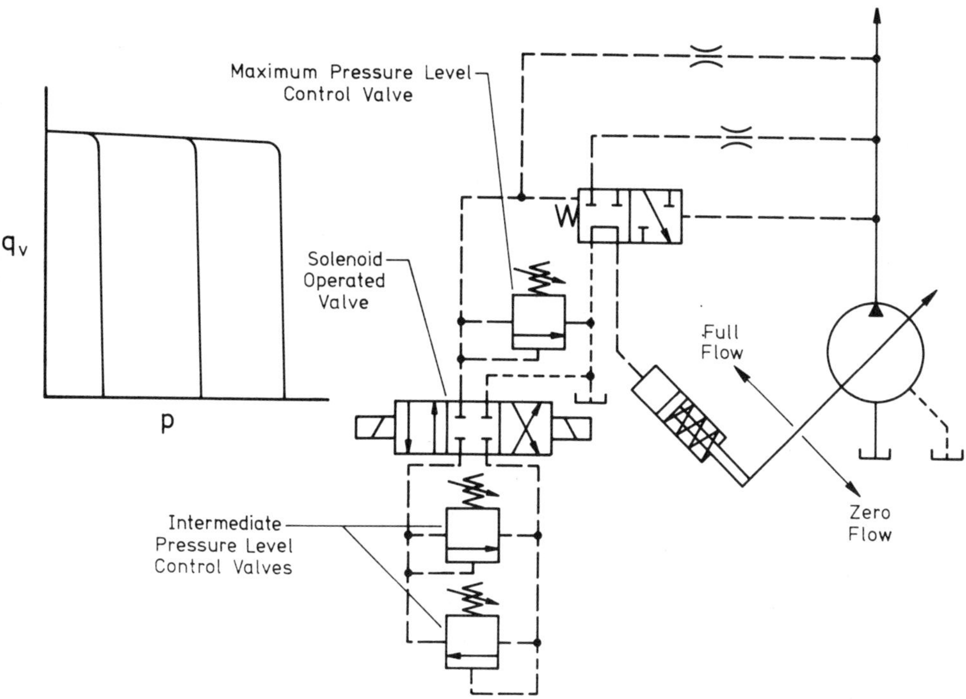

FIGURE 4-28 Triple Pressure Control

hydraulic connections can be packaged readily with the pump as a self-contained unit.

X. FIXED DISPLACEMENT PUMP CONTROLS

There are some types of auxiliary system valves that broaden the utility of fixed displacement pumps and are in sufficiently frequent demand to warrant packaging them as part of the pump assembly. Alternately, they are sometimes available as compact, integrated assemblies that can be attached directly to the pump outlet port.

A. Flow-Control Valve

Preceding Section IX-E described a means for obtaining a constant flow output from a variable-displacement pump over a range of operating speeds. This same type of requirement is often associated with fixed displacement pumps as well,

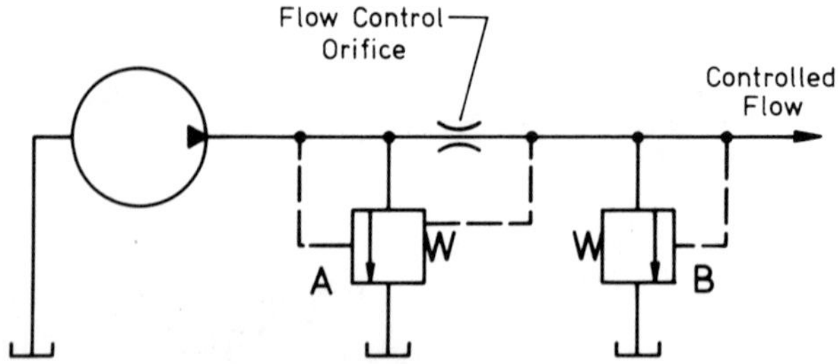

FIGURE 4-29 Fixed Displacement Pump Flow Control Valve Assembly

particularly when used for power steering or in other systems where a constant flow rate is desired regardless of pump rpm.

Figure 4-29 schematically illustrates constant pump outlet flow. The flow through the flow control orifice produces a pressure drop which acts across valve A. When the pressure drop reaches the value of the spring setting of that valve it opens, diverting the excess flow to the reservoir. Valve B acts as a safety valve for the system, limiting the maximum pressure that can be applied to the load.

A characteristic control regulation curve is shown in Figure 30. There is a small increase in controlled flow as pump rpm increases due to the regulating characteristics of valve A. The magnitude is usually small–around 20%. The value of controlled flow is, of course, established by selection of the flow control orifice size and by the spring setting of valve A. The flow control system

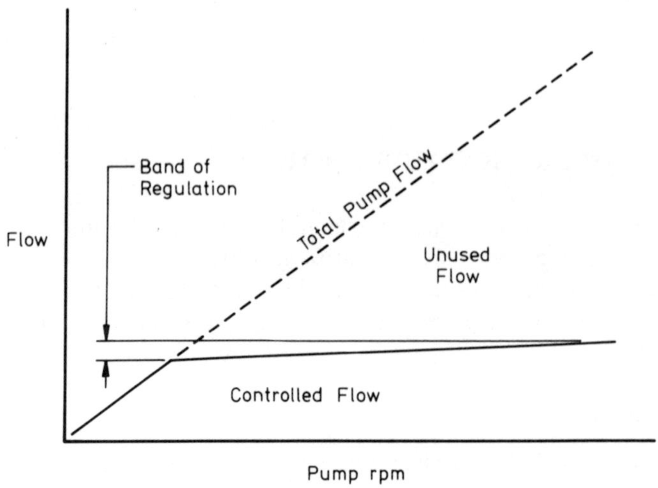

FIGURE 4-30 Flow Control Valve Regulation Curve

power loss at any pump rpm is proportional to that spring setting (i.e., the pressure drop across the flow control orifice) multiplied by the controlled flow plus the pump operating pressure multiplied by the unused flow. This last is the power loss across valve A and is directly related to the pressure required by the load. When the pressure is normally low with only occasional peaks, as in power steering, the overall flow control losses are also low.

B. Priority Valve

In some applications where the controlled constant pump flow is used to serve a critical function such as steering, there may be other secondary functions that could make use of the available unused flow. As Figure 4-30 illustrates, at higher pump speeds, the available flow often greatly exceeds the controlled value and these higher speeds may represent the normal operating mode of the application.

The valve arrangement of Figure 4-31 enables this excess flow to be used, rather than dumped to the reservoir, while still maintaining the guaranteed minimum flow for the primary function. Again, valve A in conjunction with the flow control orifice establishes the controlled primary flow. The remainder, instead of bypassing to the reservoir, is directed to the secondary load or loads. In this priority system there is an added complicating factor: The secondary pressure demand may be higher than that of the primary and this higher pressure is reflected back to the pump. This added pressure will increase the pressure acting across the flow control orifice which, in turn, will tend to increase the primary flow.

To prevent this, valve A has an added function. When the secondary pressure exceeds the primary pressure, valve A shifts further and begins to close a variable orifice that is in series with the flow control orifice. This change compensates for the added pump pressure and holds the primary flow at a nearly constant level. Figure 4-32 shows the effect of operating with the primary and

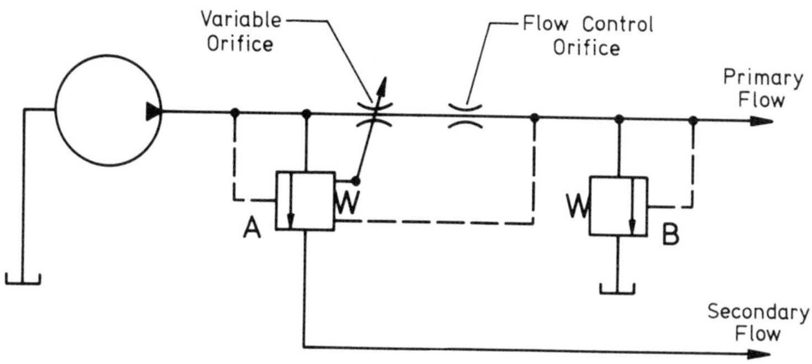

FIGURE 4-31 Fixed Displacement Pump Priority Valve

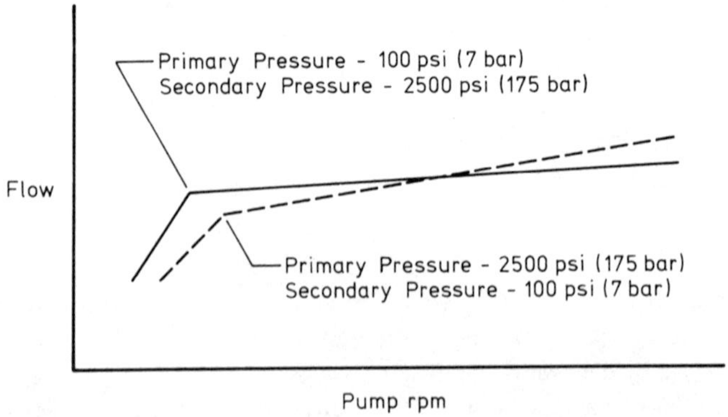

FIGURE 4-32 Priority Valve Regulation Curve

secondary flows alternately at their extreme limits. The influence is seen to be small.

In the priority valve assembly, valve B protects only the primary flow from excessive pressure. The secondary circuit must contain its own relief valve.

It is evident that the power loss in a priority valve circuit can be relatively high when the primary pressure is high and the secondary pressure low as in the flow control circuit. Power loss can also be relatively high when the opposite condition exists because of the high-pressure drop that will occur across the combined variable and fixed flow control orifices. These extremes are usually found only intermittently and for short durations when priority circuits are used. If continuous high-pressure demands occur in either circuit in a given application, some form of dual pump circuit is usually used to conserve energy and avoid system overheating.

C. Dual-Pump Unloading Valve

The use of pressure-sensitive valves for unloading one or more pumps in a multi-pump circuit was described in Sections IV and VIII-E of this chapter (cf., Figures 4-4 and 4-21). In systems where the pumps operate over a wide range of speeds, it may be desirable to utilize an unloading valve that is flow-sensitive rather than pressure-sensitive. Figure 4-33 shows such a valve. It is sensitive to the pressure drop through a flow control orifice. When the spring setting of the valve is reached, it begins to shift, bleeding off some of the flow of the second pump to drain. When the pump rpm reaches a point where the first pump can satisfy the flow control setting, the valve is fully shifted and all of the output of the second pump is unloaded to drain.

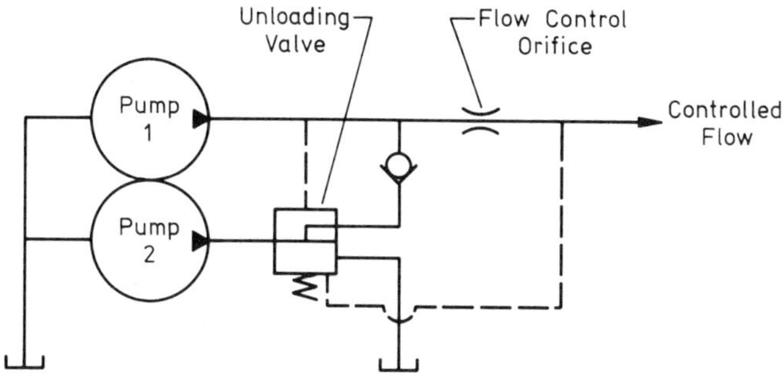

FIGURE 4-33 Dual Pump Flow-Sensitive Unloading Valve

The resulting flow curve is shown in Figure 4-34. At point A, the unloading valve begins to shift and at point B it is completely shifted. Between the two points, the metering characteristics of the unloading valve maintain an essentially constant flow. Beyond point B, the first pump alone supplies system flow.

During the period when the second pump is metering part of its flow to drain, the power loss of this system is completely analogous to that of the single pump flow control valve in Section A and Figure 4-29, above. When the valve is fully shifted, however, the complete unloading of the second pump brings the total power consumption to a minimum value. Note, however, that unlike the

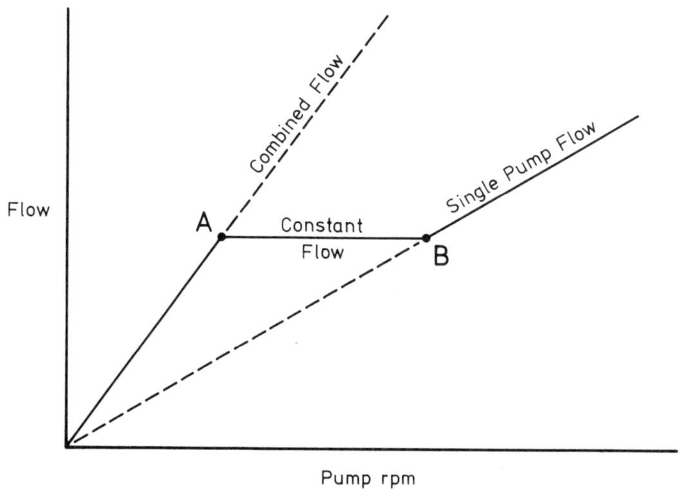

FIGURE 4-34 Dual Pump Unloading Valve Flow Curve

flow control valve of Section A, the flow beyond point B is no longer constant but rises with rpm.

D. Dual Pump Diverter Valve

In the schematic of Figure 4-33, it is evident that the flow diverted from the main system by the unloading valve can be supplied to a second system instead of being directed to drain. When this is done, flow to the secondary system begins when combined pump flow reaches point A of Figure 4-34. At point B, the entire flow of the second pump is available to the second system.

This type of circuit is most frequently used to assure good power steering response at engine idle, minimum power loss at higher engine rpm, and the ability to operate an auxiliary load when desired.

5
Hydraulic Motors

I. INTRODUCTION

The popular concept of a hydraulic motor is that it is nothing but a pump run backwards. This may be superficially true, but there are many differences in operating demands between pumps and motors. Because of this, a design that is completely acceptable as a pump may operate poorly as a motor in certain kinds of applications. Many motor designs have internal design features differing from those found in the corresponding types of pumps; in fact, some motors have no pump counterpart.

Some of the operating demands that differ between pumps and motors:

Pumps	*Motors*
Most operators view a pump as a means for delivering flow at a given pressure. There is an emphasis on volumetric efficiency.	Most operators view a motor as a means for delivering torque at a given pressure. There is an emphasis on torque (i.e., mechanical) efficiency.
Pumps usually operate at a high and relatively constant rpm.	Motors operate over a wide range

Pumps are usually expected to deliver high pressure at normal (i.e., high) rpm.

of speeds, and may spend considerable time at very low rpm.

Motors often encounter the highest pressure at zero or very low rpm.

Pumps usually rotate in only one direction in a given application. Except in transmission pumps, the direction of flow and of pressure application remains unchanged.

Most motors are required to reverse their direction of rotation. Many are also required to operate in a pumping mode for braking (overrunning) loads.

Pumps operate continuously in most systems and encounter relatively slow changes in fluid temperature.

Motors may be idle for long periods. A cold motor may be subject to frequent thermal shocks when operated.

The majority of pumps are pad-mounted and the drive shafts are not subject to side loads.

Many motors receive high shaft side-loads from pulleys, sprockets, or gears, or when used as direct wheel mounts.

The choice of motor type can be influenced strongly by the application considerations outlined above. The following review of typical motor types will serve as a means for initially narrowing the choice for any given application. Because similar designs from different manufacturers may have somewhat different characteristics, however, the manufacturers' literature should be consulted for a final selection.

In general, a motor performs the function of converting fluid power to rotary power or converting fluid pressure to torque. Torque is a function of pressure or, more properly, the motor input pressure level is determined by the resisting torque.

In the simple hydraulic circuit diagram of Figure 5-1, displacement of the directional control valve allows hydraulic flow to be metered to the motor. The resulting pressure differential, ΔP, across the motor is a direct function of the resisting torque. Motor speed, in turn, is directly related to the metering opening of the valve and the difference between the pump output pressure and the motor ΔP. If the resisting torque increases, the motor ΔP automatically increases. If the control valve opening is held constant, the motor rpm will be reduced because the reduced pressure difference between pump output pressure and load pressure produces less flow through the valve.

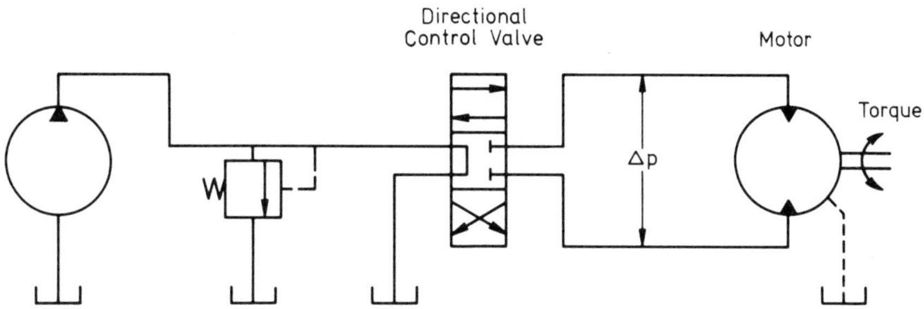

FIGURE 5-1 Hydraulic Motor Circuit Diagram

Hydraulic motors can be stalled without damage. It should be noted, however, that in circuits such as Figure 5-1, operation in a stalled mode implies continuous pumping over the system relief valve which can lead to system overheating.

While continuous pressurization of a motor in a stalled position (as in winches or hoists) is not damaging, extremely rapid stops or starts should be avoided. The torque to inertia ratio of hydraulic motors is extremely high—particularly in high speed motors—and it is possible for the motor to reach full operating speed within a few milliseconds. This rate of acceleration can produce shock loads somewhere along the mechanical drive train which in turn are reflected back to the motor.

II. HIGH-SPEED MOTORS

In many applications, the motor operates continuously at relatively high rpm. Examples are fan drives, generator drives, and compressor drives. While the speed is high and reasonably constant, the load may be either steady, as in a fan drive, or quite variable, as in compressors or generators.

The types of motors used for high-speed operation have direct counterparts in the family of fixed displacement pumps. While some details of internal construction may be modified to favor bidirectional operation or other special motoring demands, the basic principles of operation remain the same. The four primary types of high-speed motors are inline piston, bent-axis piston, vane and gear.

A. Inline Piston Motors

The inline piston motor offers a favorable combination of high efficiency, ruggedness and relative simplicity. It is available both as a fixed displacement

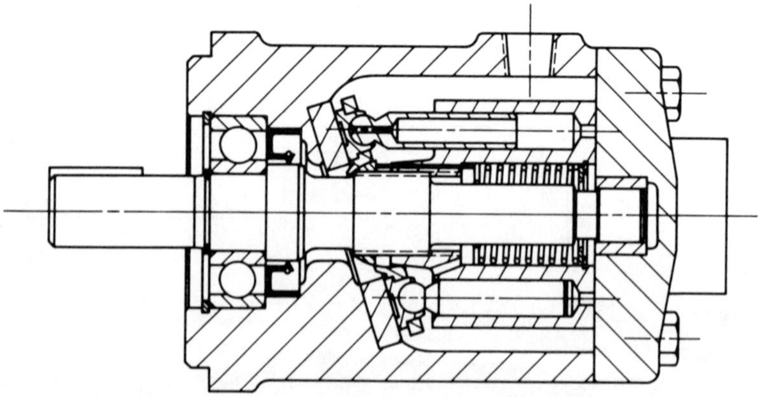

FIGURE 5-2 Fixed Displacement Inline Piston Motor

design, as shown in Figure 5-2, or in the variable displacement version shown in Figure 5-3. The mechanism used to allow varying the displacement–a variable angle yoke–is a direct counterpart of the mechanism used in the variable displacement pump.

In a motor, the torque delivered at a given pressure input is proportional to the displacement. The motor speed for a given input flow is inversely proportional to the displacement. In other words, reducing the displacement of a variable displacement motor makes torque output go down but speed go up. This allows a variable displacement motor to drive heavy loads at a low speed while still maintaining high-speed capability when the load is lighter.

The high efficiency of the inline piston motor and its adaptability to high-pressure systems makes it a popular choice for many industrial and mobile systems. The variable displacement capability in addition makes it particularly

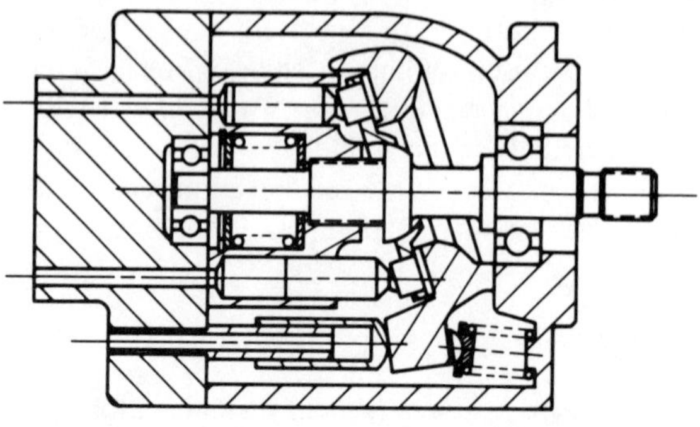

FIGURE 5-3 Variable Displacement Inline Piston Motor

suitable for industrial machines that require a rapid traverse operation at low load followed by slower, high-load operation. A comparable example in mobile vehicles is in wheel drives. A low-speed, high-torque capability is often needed for high-load operation on rough terrain, and a higher speed at lighter load for operation on paved roads.

1. Efficiency The typical efficiency pattern of a small, medium-duty inline piston motor is shown in Figure 5-4. Note that the overall efficiency is above 90% over an extremely wide range of speeds and pressures. The same is true of the larger, heavy-duty inline piston motor of Figure 5-5. Optimum performance of that unit occurs near mid-range for both speed and pressure where the motor is likely to see the greatest share of its operation. The variation in performance over the remainder of the range is not large.

The two performance curves of Figures 5-4 and 5-5 cover units having a range of nearly 10 to 1 in displacement and more than 10 to 1 in torque and power capability. Notice the efficiencies are similar. This is typical of inline piston units which have uniformly high volumetric efficiency and moderate internal friction losses, regardless of size.

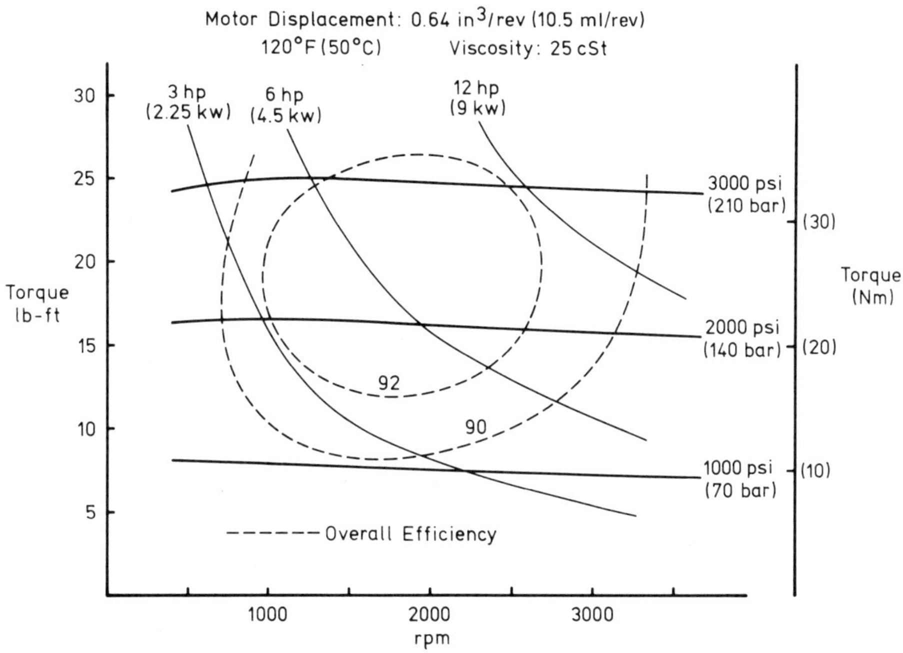

FIGURE 5-4 Medium-Duty Inline Piston Motor Efficiency

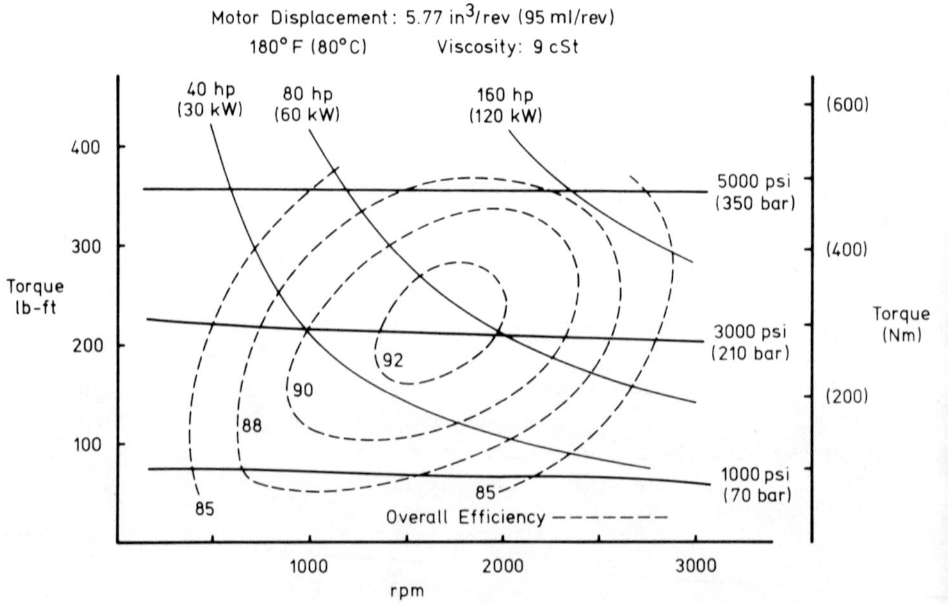

FIGURE 5-5 Heavy-Duty Inline Piston Motor Efficiency

2. Operational Features The high volumetric efficiency of inline piston motors permits them to produce steady speed, at a given flow, even though the torque demands may vary. They also have excellent braking or overrunning load capability. This feature is important in wheel drives, swing drives, winches and many other uses.

High volumetric efficiency also makes inline motor performance largely insensitive to the changes in viscosity that occur at higher operating temperatures. Efficiency remains virtually unaffected over a wide temperature range.

As in inline pumps, inline piston motor shafts are bearing-mounted. This gives good sideload capability for pulley, gear, or sprocket drives.

In variable displacement versions, the piston stroke position may be varied manually or by hydraulic actuators which permit remote control. When used for wheel drives or many other uses, direct mechanical connections may be difficult. The simplest hydraulic control employs a two-position valve which will position the motor either in full displacement or at a second lower stroke, usually around 50% of maximum. This allows a 2:1 change in torque and speed, depending on operating demands.

Some systems employ more sophisticated types of motor stroke controls. These sense motor load conditions by sensing motor-inlet pressure and adjust motor stroke accordingly; the control calls for full stroke when the load is high and adjusts stroke downward for higher speed as the load reduces. This type of automatic control will be discussed at the end of the chapter together with other types of motor controls.

Inline piston motors, like their pump counterparts, require a case drain connection to carry leakage fluid from the motor case to the reservoir. An alternate in unidirectional applications is to connect the case drain to the motor outlet line. With either connection, care must be taken to guard against case drain flow restrictions or back pressure surges which could damage the shaft seal.

B. Bent-Axis Piston Motors

The bent-axis piston motor is a close counterpart of the inline motor in efficiency and in overall capability. Its general construction, shown in Figure 5-6, is identical to the construction of a bent-axis pump. While somewhat more complicated in construction than an inline piston motor, it is capable of high efficiency and can be designed for heavy duty operation.

1. Efficiency Figure 5-7 shows the efficiency pattern of a small heavy-duty bent-axis piston motor. Because of its high volumetric efficiency, as in the inline motor, the high efficiency holds over an exceptionally broad range of speeds and pressures. This efficiency pattern is typical of the bent-axis design over a wide range of displacements.

2. Operational Features Operational features parallel those of the inline motor design:

High, volumetric efficiency produces steady speed, at a given flow, even though torque demand may vary.
Braking and overrunning load capability is excellent.

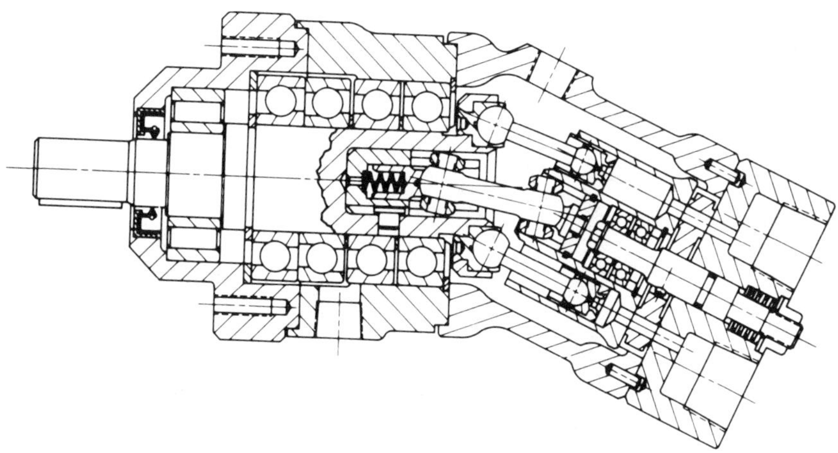

FIGURE 5-6 Bent-Axis Motor Construction

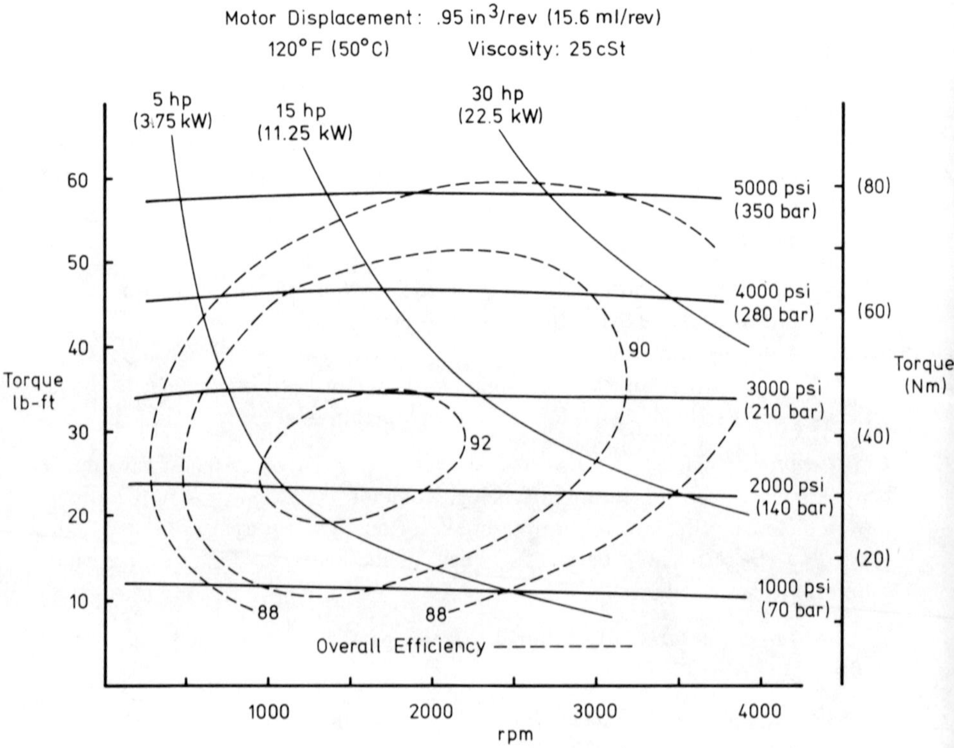

FIGURE 5-7 Bent-Axis Motor Efficiency

Shaft bearings permit good sideload capability for gear, pulley, or sprocket drives.

Variable displacement motors are available in some model series.

Bent-axis motors require a case drain connection and provisions must be made to avoid excessive back pressure which could damage the shaft seal.

C. High-Speed Balanced Vane Motors

The key characteristics of the balanced vane motor are its compact size, its simple construction, and the absence of shaft-bearing loads, all of which lead to high reliability. The balanced vane motor construction is closely similar to that of the balanced vane pump, with the design differing only in detail to permit birotational operation. Also, in vane motors, the vanes are spring-loaded against the cam ring to ensure contact before pressure is applied.

The balanced vane motor has wide usage in both industrial and mobile applications requiring a general purpose motor for medium-duty service. The housing of the pumping mechanism in a readily replaceable cartridge also adds to its utility.

1. Efficiency The efficiency of high-speed vane motors is quite favorable when they are used at their optimum power level. Figure 5-8 is a typical efficiency plot showing that performance tends to peak near the rated level of speed and pressure. The balanced motor design, because it has no bearing loads, is capable of continuous operation near full-rated conditions with high reliability.

Operation at low speeds is less favorable because of the relatively high volumetric losses compared to piston motors. The same is true of high temperature operation where the viscosity is low and leakage increases. Regardless of these theoretical limitations, the balanced vane motor is an excellent choice for many applications requiring steady, reliable duty with good efficiency near rated operating conditions.

2. Operational Features High-speed balanced vane motors have braking and overrunning load capability at high or moderate speeds. Relatively low volumetric efficiency imposes some limitations in braking capability at very low speeds, however.

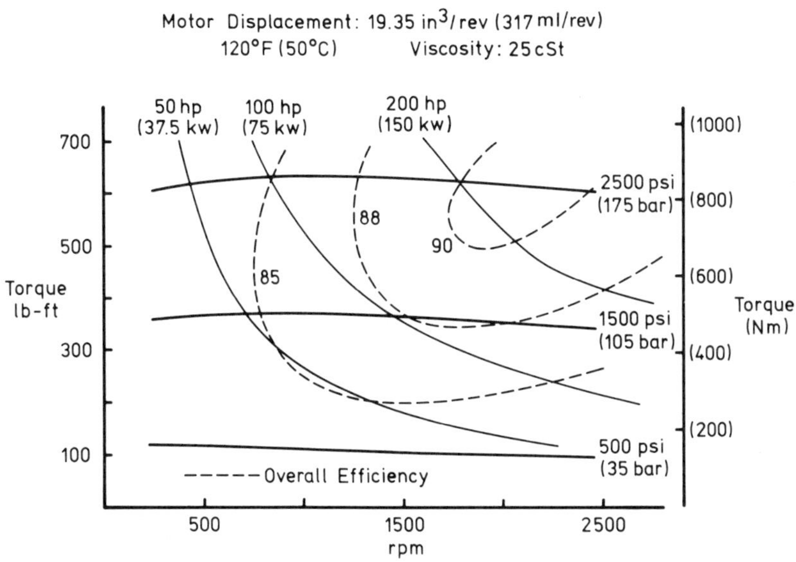

FIGURE 5-8 High-Speed Balanced Vane Motor Efficiency

Ability to accelerate from zero rpm under load is very good. Durability is not affected by continuous operation at low speed under load. Continuous smooth operation at the lower end of the speed range below 100 or 200 rpm may or may not be practical, however, depending upon the characteristics of the load being driven. High inertia of the load coupled with moderate and uniform torque demands facilitate smooth operation at lower speeds.

The presence of an antifriction bearing at the shaft end allows good side load capability for gear, pulley or sprocket drives.

Vane motors, like piston motors, require a case drain connection to vent the shaft seal area to the reservoir.

D. Gear Motors

Gear motors are compact and simple in design. Most gear motors are direct outgrowths of gear pump designs, with internal features modified to allow birotational operation. They are suitable for medium duty general utility operation.

1. Efficiency and Operating Characteristics Typical gear motor designs have moderate and reasonably uniform efficiency over a broad range of speeds and pressures. Efficiency of a typical motor is shown in Figure 5-9. Gear motors are

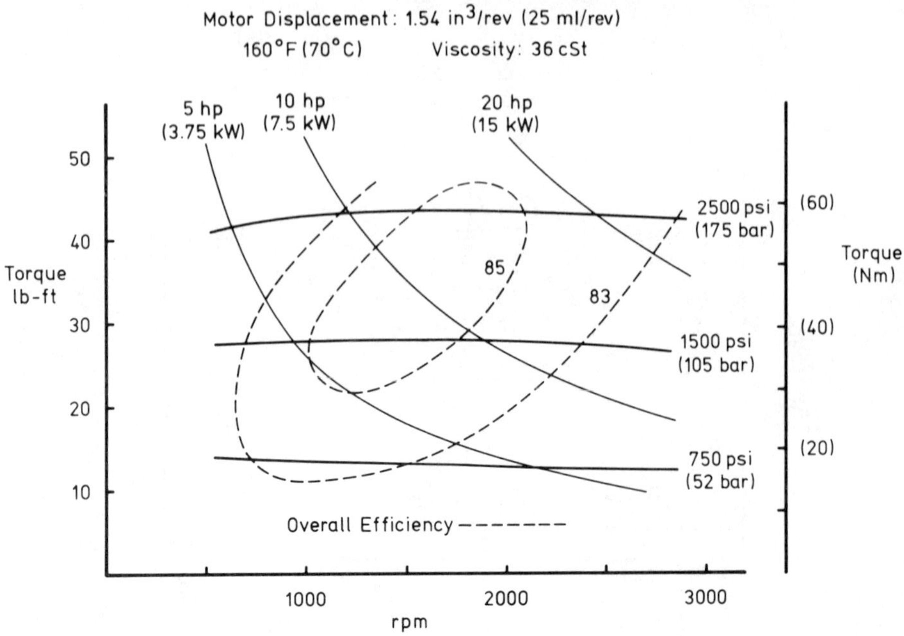

FIGURE 5-9 Gear Motor Efficiency

capable of continuous operation near rated power levels with good life and reliability.

Because the bearings in gear motors are loaded and depend upon the generation of hydrostatic films, there are limitations in operation in the lower 20% or so of the speed range when operating pressure is high. In some gear motor designs, needle bearings are used in place of journal bearings to avoid this limitation.

The loaded bearings in gear motors also tend to limit drive shaft side load capability for pulley or sprocket drives. Gear motors, like piston and vane motors, require a case drain connection to vent the shaft seal area to the reservoir.

III. HIGH-TORQUE, LOW-SPEED DRIVES

There are frequent demands by machines or vehicles for high-torque output at low rpm. A typical example is found in wheel or track drives on heavy construction vehicles. The need for high torque is also often accompanied by the need to accommodate high sideloads from sprockets or direct wheel mounts.

One solution to this application problem has been the design of planetary drives to convert the high input speed of high speed motors to the very low speed and high torque demanded by the vehicle. These drives are also designed to accept the high sideloads that are characteristic of these applications.

A typical high ratio reduction gear assembly is shown in Figure 5-10. It is driven by a heavy-duty, inline piston motor. There is an initial small ratio change through a high-speed bevel gear/spur gear combination. The final large reduction is through a double planetary gear. The output shaft is mounted on heavy-duty bearings capable of accepting substantial side loads.

This type of high-torque, low-speed drive has been designed with output torque capabilities as high as 200,000 lb-ft (270,000 Nm) and with overall gear ratios as high as 200 to 1. With this gear ratio, output speed is about 10 or 15 rpm.

When used as a track drive on a vehicle working on rough terrain, there is a premium on keeping the profile of the drive compact so as to avoid possible impact damage. One design which accomplishes this is shown in Figure 5-11. It embodies a built-in inline piston motor and a planetary reduction gear with the output shaft mounted on heavy-duty bearings.

Rated output torque by the assembly shown is 5000 lb-ft (6800 Nm) at an output speed of 100 rpm. Maximum diameter is only 14 in (355 mm) and the projection of the motor housing above the mounting flange is well under that dimension.

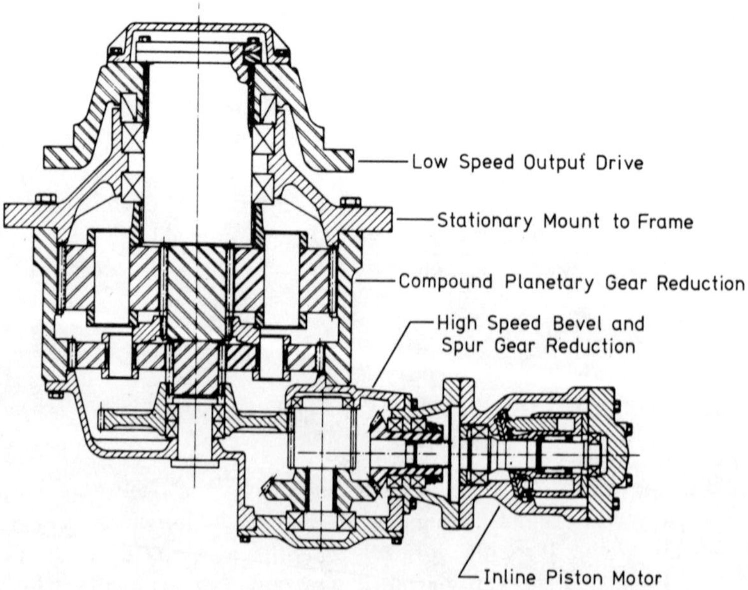

FIGURE 5-10 Planetary High-Torque Low-Speed Drive

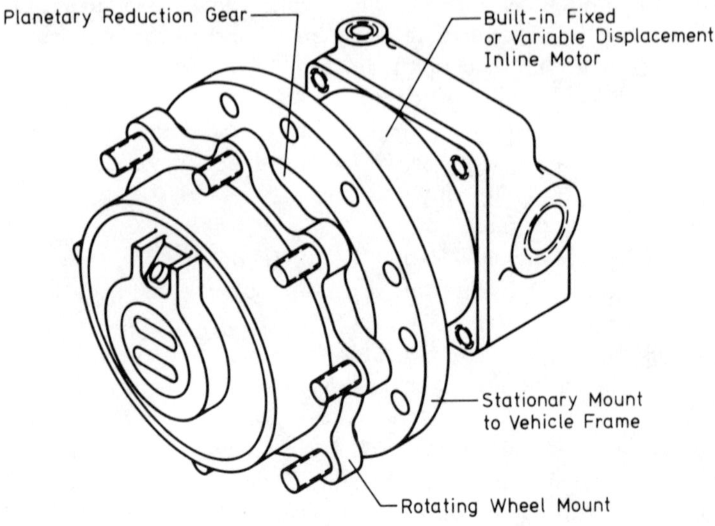

FIGURE 5-11 Planetary High-Torque Drive with Integral Hydraulic Motor

The housing of this drive is designed to accommodate either a fixed displacement or a two position variable displacement motor. When the variable unit is used, it provides the choice of two output speed and torque levels. Other features available with these types of drives are optional parking or dynamic brakes and a free-wheeling disconnect.

IV. MEDIUM- AND HEAVY-DUTY HIGH-TORQUE, LOW-SPEED MOTORS

An increasingly popular source of high-torque, low-speed output is found in the various high-torque motor designs that have been developed. They are designed specifically for efficient motoring operation and have no direct counterparts in the fields of either pumps or high-speed motors. They are simple to install and operate, and require no special maintenance since all parts are self-lubricated by the operating fluid medium.

All designs are capable of carrying overrunning loads and braking to very low rpm. They are characteristically quiet in operation, and life under constant operating loads is excellent.

The four designs covered in the following sections typify four well thought out approaches to high-torque, low-speed motor design. Each has interesting features and is capable of providing efficient, durable service.

A. Radial Piston/Crankshaft Design with Bearing-Supported Pistons

The design illustrated in Figure 5-12 originated with Bignozzi in Italy. It is a 7-piston design with the cylindrical shoes of the pistons bearing directly against the outer race of a heavy-duty, antifriction bearing. Relative motion between the piston shoes and crankshaft is carried by the bearing, reducing rotational friction and preventing shoe wear.

Fluid is distributed to the pistons through a rotating transfer gland having face contact. The transfer gland is pressure-loaded to minimize leakage losses at the contact face and accommodate thermal expansion and contraction of the rotating parts.

The crankshaft is carried on heavy-duty, antifriction bearings. While these are loaded by the piston forces, there is excess capacity available to carry side-loads on the shaft.

Typical performance of the Bignozzi motor is illustrated in Figure 5-13. The maximum level of pressure shown in that figure is considered to represent intermittent peak load operation. The normal operating range would be expected to fall between 1500 and 4500 psi (105 to 315 bar) where overall efficiency is from 86 to 92%. These types of motors require a case drain connection to carry away leakage fluid and prevent excessive pressure against the shaft seal.

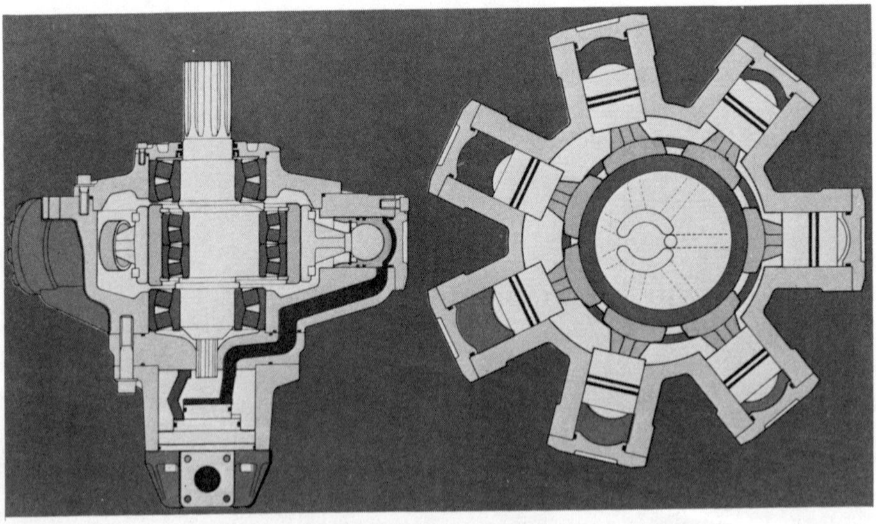

FIGURE 5-12 Radial Piston/Crankshaft Motor with Bearing-Supported Pistons (Cross-section drawing provided by Bignozzi, Italy, and National Hydraulics Co., Ltd., Concord, Ontario, Canada.)

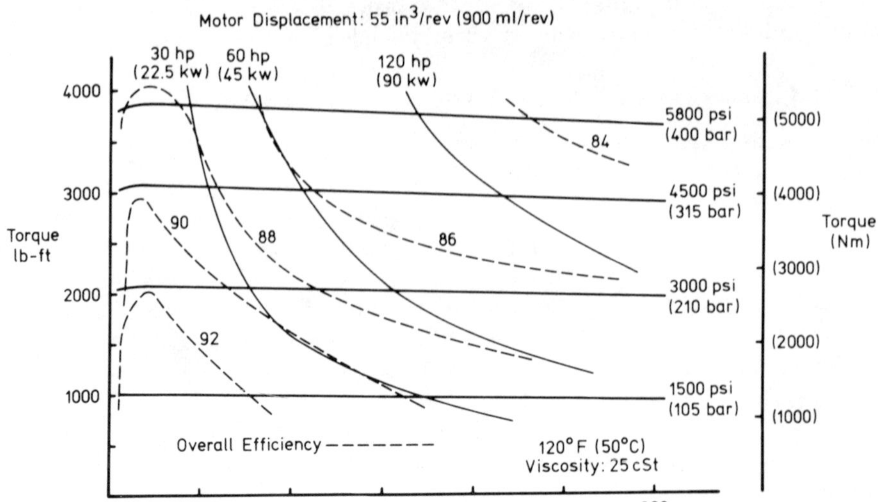

FIGURE 5-13 Bignozzi Motor Performance (Performance curves based on data supplied by Bignozzi, Italy, and National Hydraulics Co., Ltd., Concord, Ontario, Canada.)

B. Radial Piston/Crankshaft Design with Hydrostatic Piston Support

The five piston radial design of Figure 5-14 originated with Riva Calzoni of Italy. It is of interesting construction. The pistons are telescoping and the bearing surface at each end is hydrostatically supported. The pistons bear against a spherical surface on the crankshaft and against spherical cylinder heads.

Fluid is distributed to the pistons through a rotating face-type transfer gland. The crankshaft is supported on heavy-duty bearings. While these bearings carry the hydraulic piston loads, there is spare capacity for shaft side loads.

Figure 5-15 shows the efficiency of a typical Calzoni motor. The 3500 psi (245 bar) pressure level is considered to be for maximum intermittent operation. Normal usage would generally fall in the 1500 to 3000 psi range (105 to 210 bar) where efficiency is 90% or more for this size motor. The Calzoni motor, like the others, requires a case drain connection to carry away the leakage fluid and avoid loading the shaft seal.

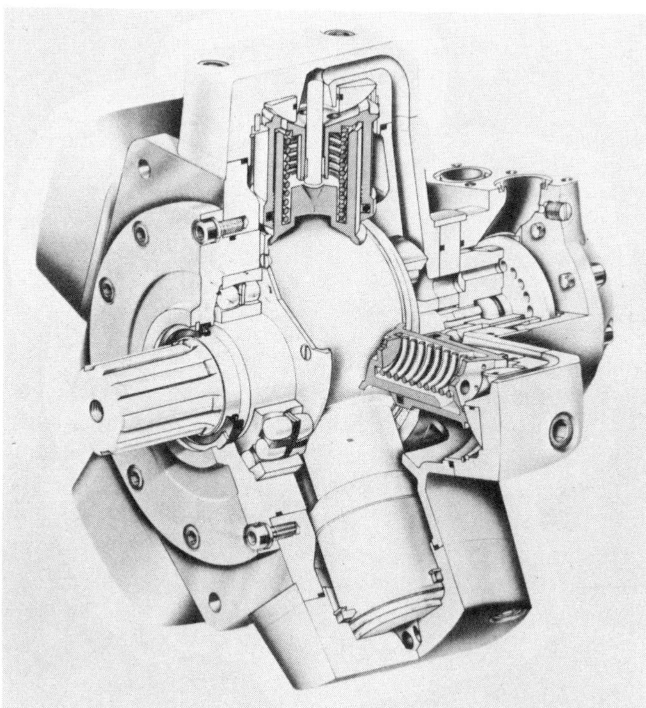

FIGURE 5-14 Radial Piston/Crankshaft Motor with Hydrostatic Piston Support (Cross-section drawing provided by Riva Calzoni, Bologna, Italy.)

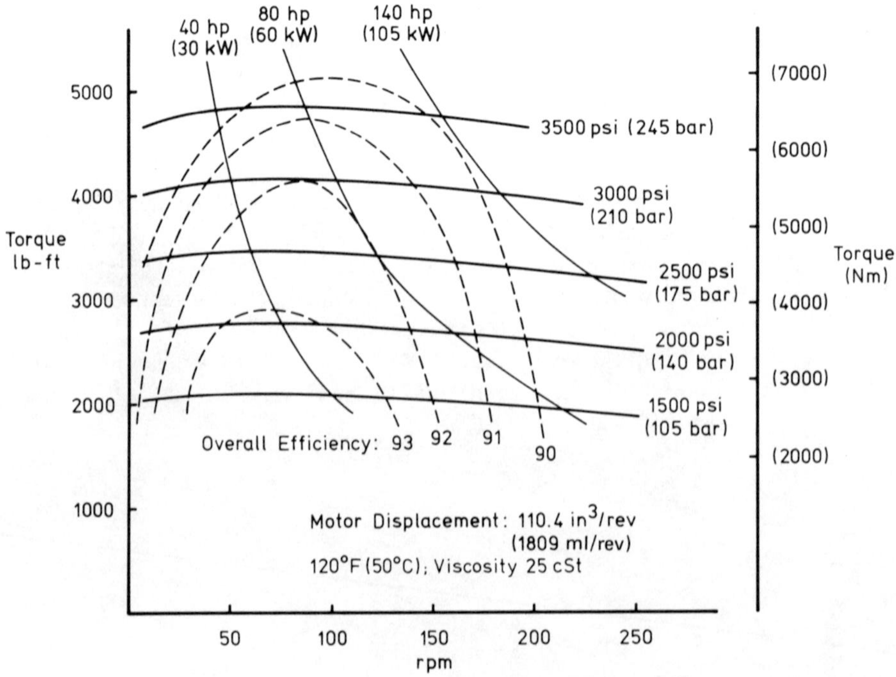

FIGURE 5-15 Calzoni Motor Performance (Performance curves based on data supplied by Riva Calzoni, Bologna, Italy.)

C. Radial Piston/Cam Design

The HM radial piston motor series, originated by Renold Limited in England, has a number of very interesting design features. As shown in Figure 5-16, the typical design consists of six radial pistons bearing against a four-lobed cam. The pistons are pressurized in opposite pairs so there is no net reaction force on the rotor. This leaves the full capacity of the heavy-duty rotor bearings available for side loads. The bearing capacity and motor housing construction permit mounting wheels directly on the motor shaft.

Fluid is ported to the pistons through a central, low leakage pintle valve. This gives the motor excellent low-speed capability and fine load-holding capacity in winches. It also guarantees positive braking against overrunning loads.

Operation is extremely smooth because of the cam design which gives completely uniform rotational motion throughout the entire 360° of rotation when there is constant input flow. There is no cogging tendency. Low-speed smoothness is further enhanced by the very low unswept volume of the fluid between the pintle valve and the pistons.

The roller in each piston is supported on a cylindrical bearing of special composition having high load-carrying capacity. The roller is aligned to the cam surface by guide rings. The pistons can be retracted to permit free-wheeling by applying low pressure in the motor case.

The overall construction of the HM motor results in a design that is very compact and light weight, as shown in Figure 5-17. It is adaptable to either open- or closed-loop systems and can accept full pressure in both ports simultaneously, as can occur with overrunning loads in a counterbalance valve system.

The combination of high volumetric efficiency and low rotational losses results in overall efficiencies that are uniformly high over a large part of the operating range, as shown in Figure 5-18. Normal operation would be considered to fall in the pressure range up to 3000 psi (210 bar) with higher pressures being available for intermittent overloads.

This motor, like the other piston units, requires a case drain connection. Case pressure must be held below motor outlet pressure unless it is desired to retract the pistons for free-wheeling operation.

D. Multi-Lobed Vane Design

A fourth type of high-torque, low-speed motor, distinctly different from the previous three in design and application, is the multi-lobed vane type. As shown in Figure 5-19, the design employs a four-lobed cam, enabling four vanes to be pressurized simultaneously. The pressure forces on the rotor are in complete radial balance, allowing the rotor to rotate freely in a shaftless assembly.

The compact form of the rotor and cam assembly allows as many as four rotating elements to be stacked together between a single pair of end bodies

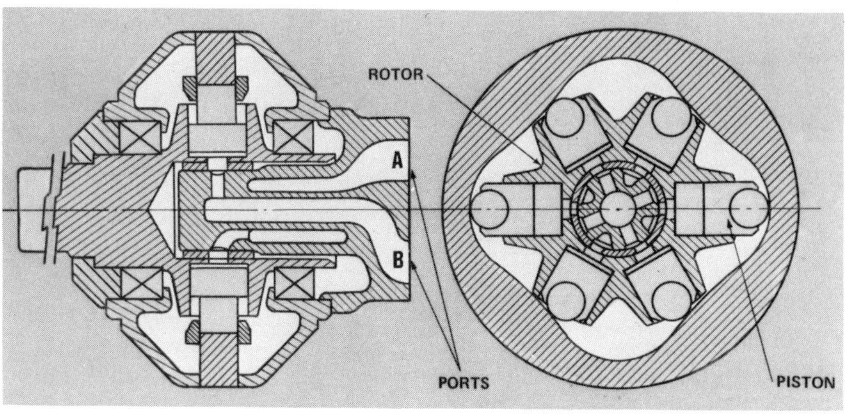

FIGURE 5-16 Radial Piston/Cam Motor (Cross-section drawing provided by Renold Ltd., Manchester, England.)

FIGURE 5-17 HM Motor Assembly (Cutaway photograph provided by Renold Ltd., Manchester, England.)

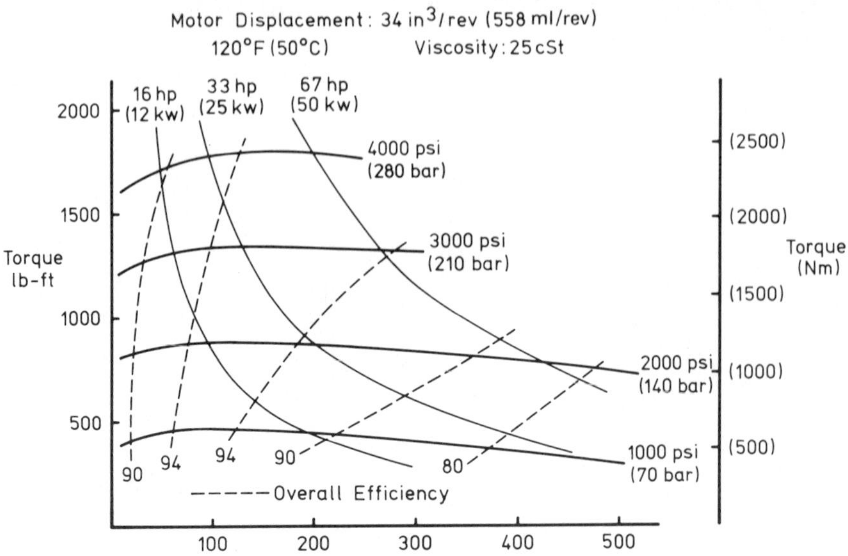

FIGURE 5-18 HM Motor Performance (Performance curves based on data supplied by Renold Ltd., Manchester, England.)

FIGURE 5-19 Multi-Lobed Vane Motor

which carry the hydraulic porting. An assembly of this type is shown in Figure 5-20. It consists of four rotors and cams, each having a torque-output capability of 5000 lb-ft (6700 Nm) at 2000 psi (140 bar) input pressure. Thus, total torque output is 20,000 lb-ft (27120 Nm).

This type of motor is widely used in plastic-molding machines where it mounts directly on the splined feed screw drive shaft. For injection of the plastic into the die, a hydraulic piston is able to move the feed screw axially through the hollow motor without interference. For other types of applications, the motor can be supplied with an integral bearing-mounted shaft.

Typical high-torque, low-speed vane motor performance is shown in Figure 5-21. Efficiency is lower than piston units due to higher volumetric losses, and can be improved by the use of higher viscosity fluids. Operation is normally in the range of 500 to 1500 or 2000 psi (35 to 105 or 140 bar) where efficiency is more nearly optimum.

An interesting feature of this motor design is that the vanes can be pressurized in individually opposed pairs if desired, rather than all four at once. When

FIGURE 5-20 Multiple-Element Vane Motor Assembly

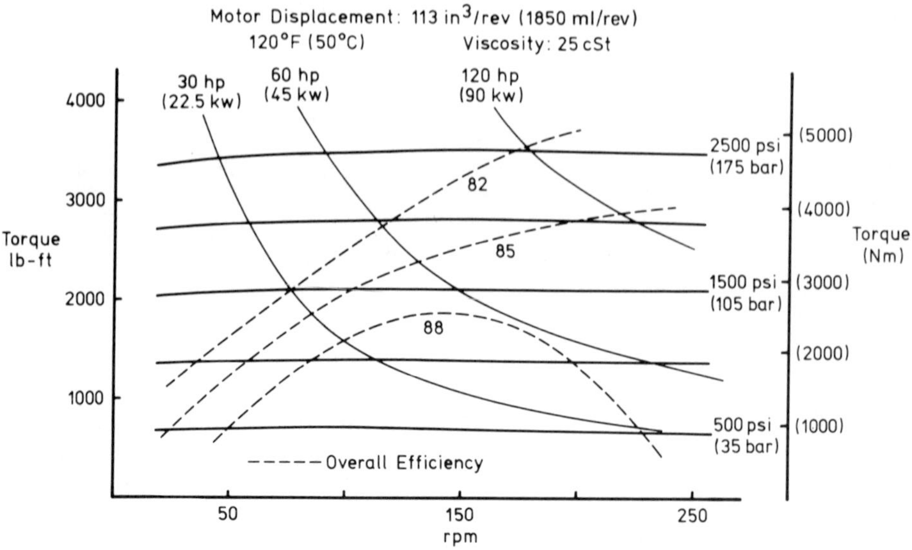

FIGURE 5-21 High-Torque, Low-Speed Vane Motor Performance

only one pair of vanes is pressurized, the displacement is reduced, thus causing a reduction in torque and increase in speed for the same pressure and flow. Since opposite vanes are pressurized together, the radial hydraulic pressure loads on the rotor remain in balance and no side loads are produced.

As a further extension of this multi-speed, multi-torque capability, an opposite pair of cam rises can be reduced in height. This reduces the area of the vane surface exposed to pressure and correspondingly reduces the displacement and the torque produced. As an example, if one opposed pair of cam rises is reduced by a third, it will have only two-thirds the displacement and deliver only two-thirds the torque of the other pair. This allows the motor to be operated at 100%, 60%, or 40% of its basic displacement, depending upon the combination of vanes that is pressurized. This feature has proved to be very useful in providing multi-speed operation for special machine operating cycles.

V. LIGHT-DUTY, HIGH-TORQUE, LOW-SPEED MOTORS

Many motor applications require only moderate torque output during the largest share of their operating life. In addition, operation may be intermittent. Momentary high-torque capability may be needed, however, to overcome randomly encountered peak loads.

The low-speed motors described in the previous section are generally too large to cover these lighter duty utility applications. The piston and vane high-torque motors have power capabilities ranging from 20 hp (15 kW) to more than 300 hp (225 kW) in some models. In contrast, there are many low-speed motor needs that fall in the range of 20 hp (15 kW) or below. They include conveyor drives, mixers, brush drives, small winches, machine tool table indexing and many others. This lower power, lighter duty range is well-covered by the orbital motor design.

A. Orbital Motors

The basic operating principle of the orbital motor is illustrated in Figure 5-22. In the design shown, the rotor has six lobes and the stator seven. The rotor and stator are closely fitted so that the rotor tips seal against the stator lobes.

If hydraulic pressure is applied in the area to the left of the centerline in Figure 5-22, it will cause the rotor to roll to the right, its center progressing in a counterclockwise direction. By admitting pressure to the correct areas as the rotor orbits within the stator, it is possible to produce continuous rotation in either direction.

When the rotor has made one complete counterclockwise orbit within the stator, the rotor tooth containing the index mark shown in Figure 5-22 will fall

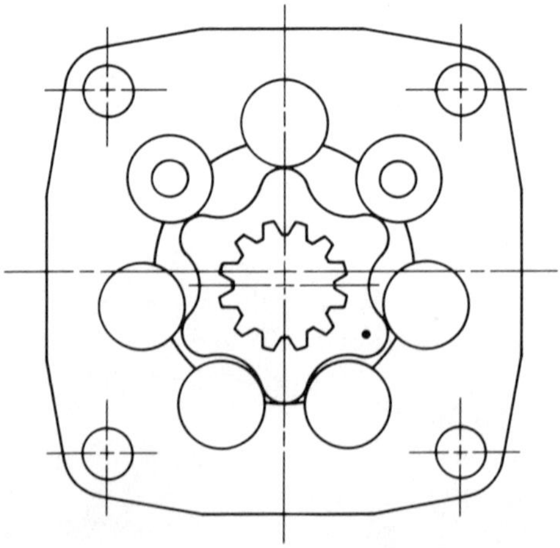

FIGURE 5-22 Orbital Motor Rotor and Stator

on the bottom location on the motor's vertical centerline. In other words, it will have indexed itself one tooth backwards in a clockwise direction. This built-in gear ratio, caused by the difference in the number of rotor and stator lobes, gives the design its torque-multiplying capability.

The indexing motion of the rotor is transmitted to the drive shaft by means of a universal link which has crowned splines to accommodate the angular misalignment. A typical arrangement of parts is shown in Figure 5-23. The orbitting

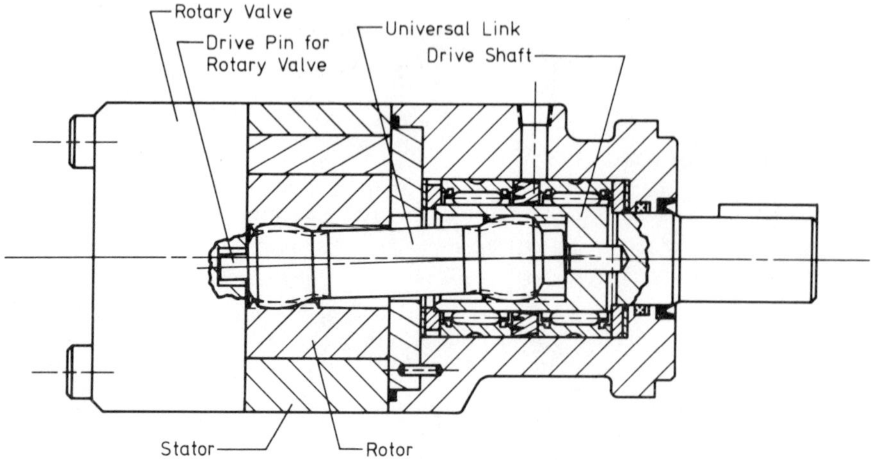

FIGURE 5-23 Orbital Motor Cross-Section

motion of the rotor and universal link is transmitted to a rotary valving mechanism which keeps pressure applied to the proper area of the rotor.

Some motors, such as the one shown in Figure 5-23 have the shaft mounted in antifriction bearings which have sufficient capacity to accept sideloading from offset drives, directly mounted brushes and the like.

The overall assembly, shown in Figure 5-24, is simple and compact and well-suited to the general run of low-power and lighter duty applications.

The orbital motors made by various manufacturers differ in many internal details and both performance and durability vary as well. Figure 5-25 can be considered to represent the upper end of the probable performance range for this type of motor, even when fluids with higher viscosity are used to minimize volumetric losses. Efficiency is not usually considered a critical issue in equipment operating at low power levels in light-duty applications, however. Emphasis is more likely to be placed on simplicity of installation, flexibility of operation, and other utilitarian considerations. In each of these respects, the orbital motor has definite advantages.

B. Two-Speed Orbital Motor

A further extension in the utility of orbital motors is offered by a two-speed version. Its basic construction is similar to the usual single-speed design, but the

FIGURE 5-24 Orbital Motor Assembly

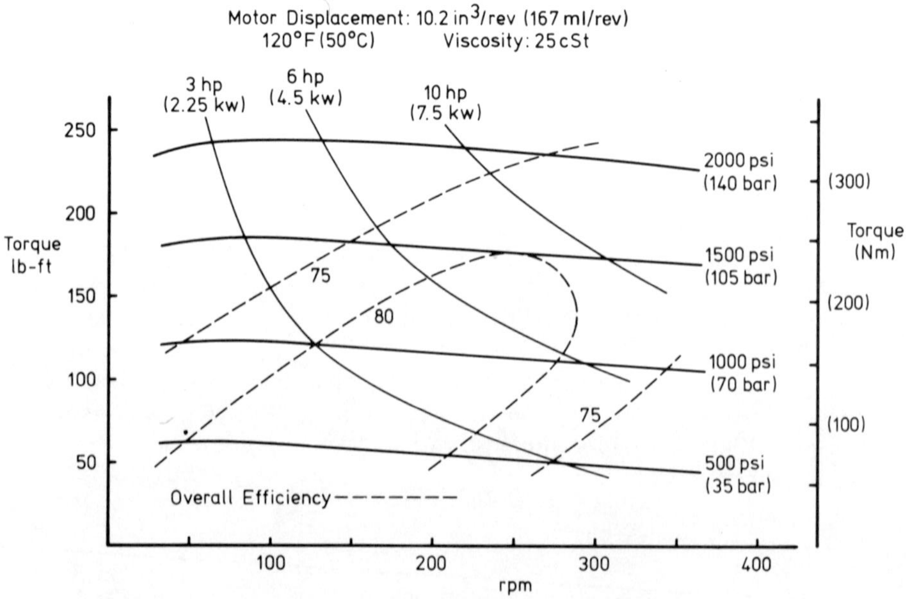

FIGURE 5-25 Orbital Motor Performance

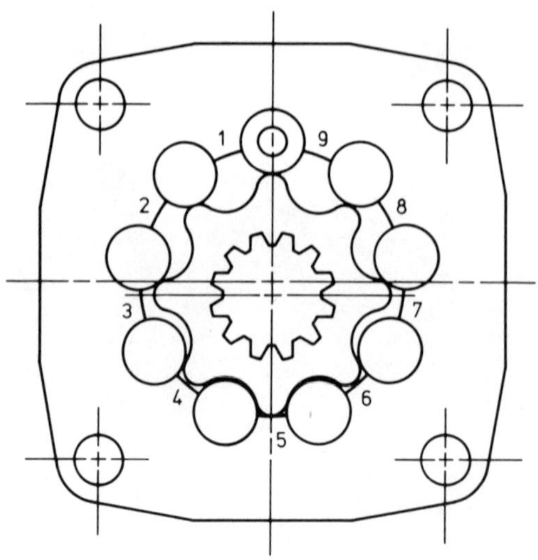

FIGURE 5-26 Two-Speed Orbital Motor

valving is so arranged that fluid pressure can be applied individually and selectively to the spaces between the rotor and stator.

Figure 5-26 shows an eight-lobed rotor within a nine-lobed stator. The points of contact between the two parts form nine individual spaces which are numbered in the figure. For full torque, high-pressure oil is fed to chambers 1, 2, 3 and 4. This provides maximum displacement of the motor and produces maximum turning effort. As the rotor orbits within the stator, pressure is successively applied to spaces 2, 3, 4 and 5, then to 3, 4, 5 and 6, etc.

When higher speed with reduced torque is desired, the pressurization in this design can be confined to chambers 1, 2 and 4. This results in a reduction in displacement and torque to two-thirds the original value while output speed is increased by 50% for the same flow. Again, as the rotor orbits, pressurization is successively switched to spaces 2, 3 and 5, etc. Opposite rotation is, of course, possible by pressurizing the opposite motor inlet port.

VI. MOTOR CONTROLS

The direction of rotation and rotational speed of hydraulic motors are normally controlled with directional control valves of the types described in Chapter 4. Either fixed displacement pumps with open-center valves or variable displacement pumps with closed-center valves may be used, depending upon system needs. These and other choices are further explored in the next chapter covering transmissions, where considerations of controllability and power conservation are discussed in some detail.

In addition to these basic system controls, there are other types of control configurations frequently encountered in hydraulic motor systems. They are often designed so that they can be manifolded directly to the motor ports and in some cases are made integral with the motor design. This close coupling improves the responsiveness of the motor by minimizing the amount of fluid under compression. In high-response systems, close coupling may in fact be essential to avoid system interactions and assure stable operation. Some of the control types require specific motor performance capabilities for proper operation. This will be discussed

A. Servomotors

Servomotors and servoactuators in general first saw wide usage in aircraft and missile control applications, then in machine tool controls. There are now also many servo-controlled functions in mobile vehicle hydraulics and, of course, in robot control systems.

Servoactuators and servomotors are used where rapid response and fineness of control are needed and also where remote operation or programmed elec-

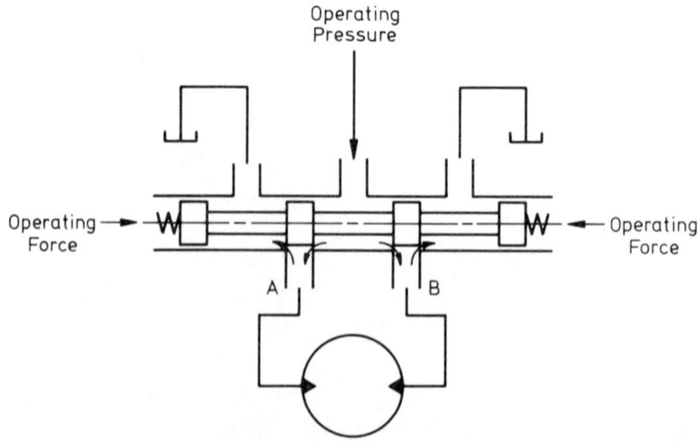

FIGURE 5-27 Hydraulic Servovalve for Motor Control

tronic control are employed. To ensure maximum response, the servovalve is generally mounted directly on the motor or actuator, keeping the amount of fluid under compression to a minimum.

1. Servovalve Operation In the typical system, the metering element of the servo valve is designed as shown in Figure 5-27, and is fed from a constant pressure source. When used with an actuator or motor with very low leakage, the hydraulic leakages across the metering edges of the valve lands produce pressures at A and B that are essentially half the system operating pressure when the valve is centered. A small motion of the valve toward the left, for example, will then cause the pressure at A to increase and that at B to decrease. This differential in pressure, when large enough to overcome the motor's internal friction and the resistance of the load, will cause the motor to rotate.

As the motor begins to rotate and there is flow past the metering edges of the valve, pressure at A will reduce and that at B will increase due to the pressure drops at the metering edges. The torque delivered by the motor and the rate of rotation will, therefore, be controlled by the amount of the valve opening.

Valve position may be controlled mechanically or by a hydraulic remote control that applies pressure to one end or the other of the spool, or by the first stage of an electrohydraulic servo valve. In systems that are computer-controlled, a feedback of position, velocity, or rpm is often employed to close the control loop for smooth, stable operation.

2. Required Motor Characteristics For smooth operation of a servo actuator or servo motor, it is important that the breakaway and running friction be low and uniform throughout the operating range. Even more important, where smooth, low-speed operation is desired, leakage must be low and uniform. These characteristics are essential to produce smooth operation under varying loads and fine resolution of position. Successful operation of a servomotor at low speed under load is completely dependent on these factors.

The need for low and uniform friction is quite obvious and the need for low and uniform leakage also becomes evident when the action of the servovalve is considered. At low motor speed, the rate of flow past the metering land of the valve is low. At low motor torque, there is little motor internal leakage and nearly all of the servo flow will be translated into motor rpm. If torque increases and motor leakage becomes high at the higher operating pressure, motor speed will be greatly reduced and the motor may, in fact, stall. This can be avoided by using a motor having high volumetric efficiency.

3. High-Speed Servomotors Suitable friction characteristics and low leakage are quite readily attainable in linear and rotary actuators but are not easily achieved in hydraulic motors. In high-speed motors, only the piston types have high enough volumetric efficiency for good control, and the bent-axis design shows a distinct advantage over the inline type for servo-operation. Depending on the torque level and inertia (i.e., flywheel) characteristics of the load, the inline motor will seldom operate smoothly below 150 rpm and rarely below 100 rpm. The bent-axis design, in contrast, will usually operate down to 50 rpm and in some applications as low as 10 rpm.

The difference in low-speed operating capability of the two motors is due to differences both in friction and in leakage. In the inline pump, the pistons are sideloaded against the cylinder bores by the angularity of the piston shoe contact on the swash plate. This side force creates undesirable friction when the pistons are stationary or only moving slowly and the lubricating film may be lost. There is also some contact force existing at the shoe face which is not completely balanced by the hydrostatic pool. The fact that all pistons are loaded to approximately half-system pressure in a servomotor when it is stationary or rotating under low-torque loads only aggravates the problem.

The bent-axis motor, on the other hand, has very low piston side forces. The angularity of the connecting rods with the cylinder bores is very small, usually only $\pm 3^\circ$. In addition, the axial thrust forces of the pistons are carried on antifriction bearings rather than on sliding shoes.

The leakage characteristics of both the inline and bent-axis designs are very good but here again the bent-axis may be slightly better. This is due primarily to the absence of the shoe face as a leakage path.

It should be remembered that the differences between the inline and bent-axis designs have a negligible effect at higher rpm's where the pistons and shoes operate on fluid films. The differences become evident only at the lower end of the speed range. They are important only when a slow, controlled speed or fine resolution of position is essential.

4. Low-Speed Servomotors The radial piston/cam and radial piston/crankshaft types of low-speed, high-torque motors generally have low-leakage and low-friction characteristics that would make them suitable for servo operation. The available sizes, however, may fall beyond the power range of most servo applications.

The radial piston/cam motor designed by Renold Limited and described in Section IV-C, is particularly well-suited for servo use. It has high volumetric efficiency and low internal friction. The cam design provides completely uniform displacement throughout the 360° of rotation, yielding constant angular velocity for constant input flow. In addition, the inertia of the rotating parts is low, permitting rapid response to input demands.

The suitability of this specific design is further enhanced by the low volume of fluid under compression, which improves system stiffness and precision of control.

B. Counterbalance Valve

When motors are used in open-loop systems and the motor can be subject to overrunning loads, it is not possible to maintain proper motor control with a simple "meter-in" system. While the metering of fluid into the motor will control motor speed when the motor is in a positive driving mode, the shift to an overrunning load will permit an uncontrolled runaway of the motor.

To permit the motor to resist an overrunning load, it is necessary to restrict the motor outlet while continuing to feed the inlet to prevent cavitation. The change from a "meter-in" to a simple "meter-out" valve is not appropriate. Full time meter-out operation unnecessarily pressurizes both the inlet and outlet of the motor simultaneously and limits motor torque output capability. The problem is resolved by the use of a counterbalance valve.

Figure 5-28 shows schematically the manner in which a counterbalance valve functions in an open-loop circuit. When the directional valve is shifted downward, fluid is directed along the upper leg of the motor circuit, past a check valve to port A of the motor. As long as there is a positive load on the motor, the pressure in that leg of the circuit maintains pressure against the upper end of the counterbalance valve. The counterbalance valve then shifts to allow free return of the fluid from port B of the motor.

If the motor encounters an overrunning load, pressure in the upper leg of the circuit will drop, allowing the counterbalance valve to return toward the centered position and restrict the motor outlet. An equilibrium position will be reached in which there is sufficient pressure in the upper leg of the circuit to maintain a positive pressure at port A and prevent cavitation while flow from port B is sufficiently restricted to prevent excessive motor speed.

The crossport relief valves shown in Figure 5-28 are often included in the circuit to limit the maximum circuit pressure and prevent any damage to circuit

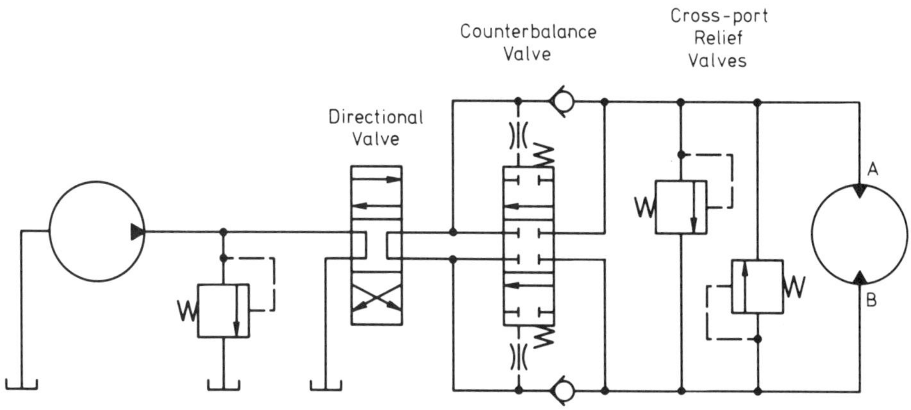

FIGURE 5-28 Counterbalance Valve

components. It should be noted, however, that when crossport relief valves are used, they should not be employed as the primary means for braking a load. In a close-coupled system, the recirculation of fluid may cause excessive temperature rise.

A specially designed control valve assembly having counterbalance capability and many other desirable features is shown manifolded directly to a motor in Figure 5-29. The motor is an HM radial piston assembly and the valve is the Vickers CMX design. It is actually a collection of valves embodied in a single housing. Together with the HM motor, it offers:

- Precise speed control with fast response;
- Remote control with a hydraulic remote control valve or electrohydraulic servovalve;
- Crossport relief valve protection to prevent damaging pressure peaks when stopping high inertia loads;
- Fail-safe load-lock provisions to hold a load if hydraulic pressure is lost; and
- A load-sensing signal for energy saving when load-sensing pumps are used.

When the special counterbalance feature is included, it prevents both cavitation and motor overheating when there is an overrunning load.

C. Free-Wheeling Valve

When a motor is used with a control valve arrangement that allows it to carry overrunning loads, both motor ports are normally blocked when the valve is in a neutral position. This is true in the circuit of Figure 5-28. It is also true when

FIGURE 5-29 HM Radial Piston Motor with CMX Control Valve

the motor is used in closed-loop transmission systems as described in the following chapter.

This blocked port condition prevents free rotation of the motor—a necessary factor in controlling overrunning loads—but is a severe handicap when a vehicle with wheels driven by the hydraulic motor needs to be towed. Free-wheeling can be provided by interposing a free-wheeling bypass valve, as shown in Figure 5-30 in the circuit to the motor. The free-wheeling valve allows free recirculation of fluid through the motor as it rotates.

In the schematic of Figure 5-28, the free-wheeling valve would be placed just downstream of the counterbalance valve. The drain connection is desirable to allow for thermal expansion and contraction of the fluid in the loop and to keep the loop filled with fluid. If towing is to be at high speed, it may be desirable to pressurize this drain connection from an external source to prevent cavitation at the motor inlet and possible motor damage.

Where the need for free-wheeling is infrequent, the free-wheeling valve is often manually actuated. For drives that are cut in and out at frequent intervals, however, actuation is usually electrical or hydraulic and remotely controlled.

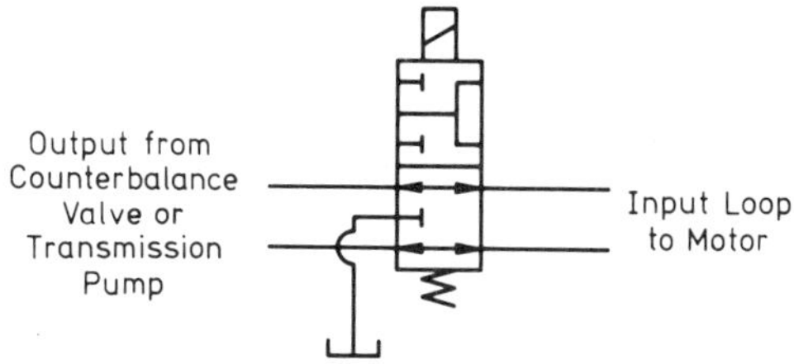

FIGURE 5-30 Free-Wheeling Bypass Valve

D. Constant-Speed Control

There is sometimes a need for very precise control of motor speed. A typical example is a hydraulically-driven emergency electrical generator, where alternating current frequency must be controlled within a narrow band. In such a system, loads can fluctuate widely and virtually instantaneously as electrical equipment is turned on or off. It is necessary for the hydraulic constant-speed control to respond rapidly and accurately to each change in power level while holding electrical frequency nearly constant.

This can be achieved by a motor speed control system of the type illustrated in Figure 5-31. A flow-sensing orifice is placed in the motor outlet line. The outlet flow of the motor is a more precise measure of motor rpm than the inlet flow, some of which is lost as internal motor leakage. The pressure drop across this flow-sensing orifice acts on a spring-balanced actuator piston, which in turn, regulates the opening of a motor inlet pressure-control valve. For accuracy of control, a piston motor with high volumetric efficiency is used—commonly a bent-axis type. The entire system is fed from a constant pressure source.

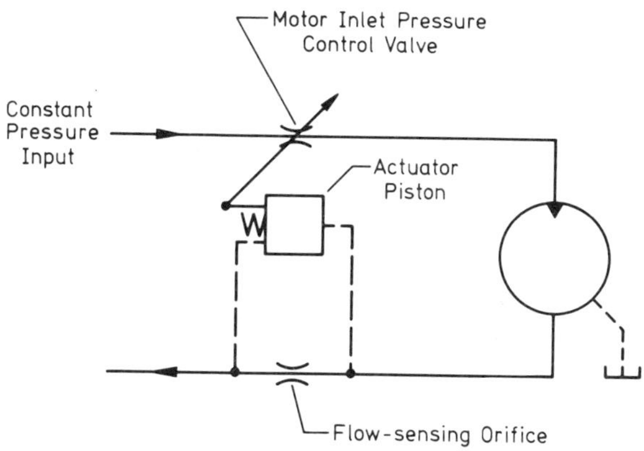

FIGURE 5-31 Constant-Speed Motor Control

To follow the functioning of the system, consider an application where the motor is operating under a steady-state torque demand. The inlet control valve will have a metering orifice opening that will throttle the system input pressure down to that required at the motor to produce the needed torque. If a higher load is supplied at the motor shaft, the motor will start to slow down. This is sensed by the orifice in the outlet line which then applies a lower pressure differential across the actuator. The actuator moves under the influence of the spring force to increase the metering opening of the inlet control valve, applying higher pressure at the motor inlet and restoring equilibrium.

Systems of this type have been capable of providing steady-state frequency control within 1%. The maximum deviation during transients is about 5%, with recovery time falling in the range of 10 to 20 milliseconds.

E. Constant-Power Control

The work output of a motor is the product of torque and speed. These values of course correspond in turn to input pressure and flow. For maximum productivity and minimum total power consumption, it is necessary for the motor to use the full input power available, regardless of variations that may occur in the load applied to the motor shaft.

A simple example of such a need is in an earth drill which may encounter varying degrees of resistance depending on soil conditions and, therefore, require varying amounts of motor torque. When a fixed displacement motor is used in this type of system, the motor output power fluctuates as the load varies and the full power capability of the system will seldom be used.

A variable displacement motor with a constant-power control offers an alternative. When torque is reduced, the motor automatically adjusts its displacement to increase its operating speed. Since motor output power is a function of torque times speed, this allows output power to remain constant and the hydraulic power available to be fully used. The added speed allows the job to be completed more quickly, improving the overall productivity of the system.

The functioning of a constant-power control is shown schematically in Figure 5-32. A pressure-compensator control senses, by means of a shuttle valve, the pressure in the hydraulic line powering the motor. If the system pressure is below the setting of the compensator, fluid is drained from the motor displacement actuator piston and motor yoke angle decreases. This allows the motor to speed up and input pressure will increase accordingly until the pressure setting of the compensator is satisfied.

If the motor then meets increased resistance, the motor input pressure will begin to rise. This causes fluid to be metered into the actuator piston to increase motor stroke until the pressure reduces to the set value of the compensator.

The effectiveness of this type of control is illustrated in Figure 5-33, which shows actual torque and speed values as compared to a theoretical constant

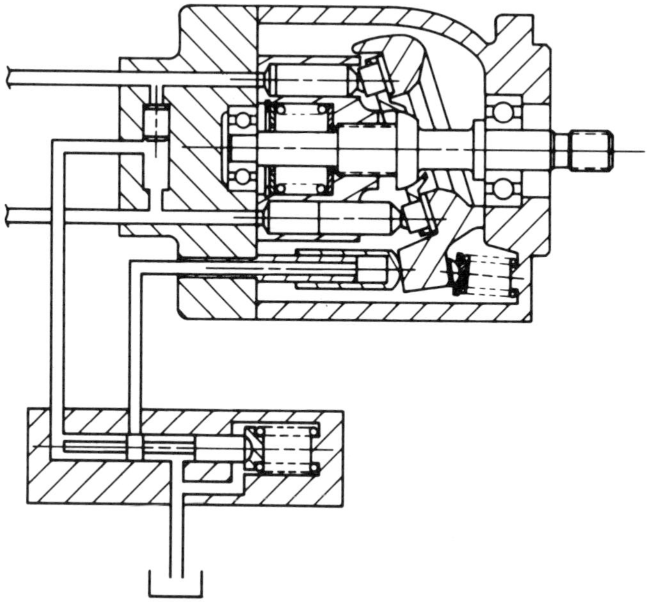

FIGURE 5-32 Variable Displacement Motor Constant-Power Control

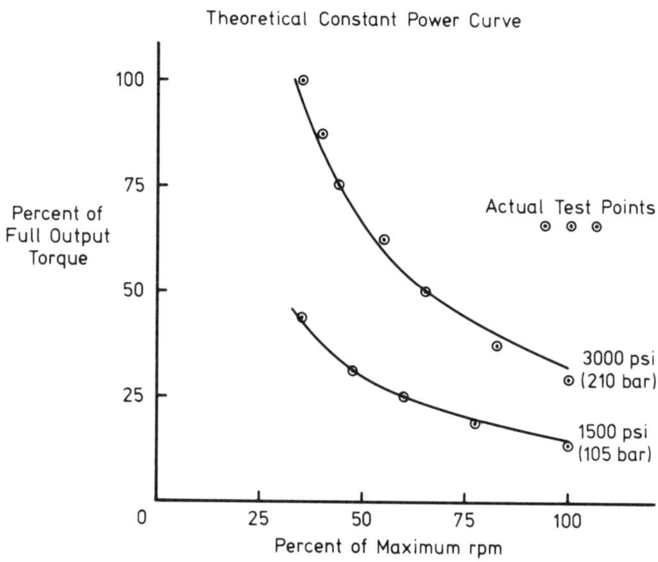

FIGURE 5-33 Constant-Power Control Torque and Speed Characteristics

power curve for two levels of compensator setting. The deviations are small—only 5% or 10%. Response time is rapid, holding any momentary pressure fluctuations to moderate values and enabling the motor to operate close to the maximum power curve at all times.

6

Hydrostatic Transmissions

I. INTRODUCTION

While many of the open-loop hydraulic circuits described in Chapter 4 can be adapted to wheel drives or other types of transmissions, this chapter describes and details closed-loop systems. In the typical system, a bidirectional motor is powered by a variable displacement pump which can be stroked "over center" to reverse the direction of flow. The primary advantages of closed-loop hydrostatic transmissions compared to mechanical transmissions are:

1. Smooth, stepless change in ratio from full-speed forward to full speed in reverse;
2. Full torque available through the entire speed range, from full forward to full reverse;
3. Full-braking (overrunning load) capability at full-torque level in both forward and reverse;
4. Simple controls with sensitive controllability for maneuvering. In skid-steer vehicles, this includes the ability to spin the vehicle about its own center;
5. Slipping of one wheel does not affect traction at the remaining wheel or wheels in drives with individual pumping sources for each wheel;

6. Motor speed relatively unaffected by load variations;
7. Load sensitivity providing excellent system efficiency. No pressure drops through control valves in the normal system;
8. Considerable installation flexibility—power can be delivered readily to remote areas;
9. Adaptable to remote control and to programmed electronic control;
10. Rapid response and ease of operator control often makes them more productive than mechanical systems; and
11. Lower cost than most mechanical systems which consist of clutch or torque converter, gear box, drive shaft, differential, etc.

Whether a closed-loop hydrostatic transmission is more or less efficient than a comparable mechanical system in terms of pure power consumption depends on the operating cycle. If extended periods are to be spent at virtually constant power with little clutch- or torque-converter slip, a mechanical system will quite probably have the edge. But in systems where the magnitude and velocity of the load are variable and some degree of maneuvering is required, the work done per kilowatt or per gallon of fuel with the hydrostatic system may well exceed that of a mechanical transmission.

This comes about because of the rapid and simple controllability of the hydrostatic system compared to a clutch/brake/gearshift/throttle system. Operator cycle time is often considerably faster with a hydrostatically controlled machine or vehicle, resulting in more work output in a given time.

When discussing transmissions, it is natural to think of wheel drives. Certainly wheel drives are typical, but there are many other uses for reversible pump/motor systems. In addition to swing drives on excavators and cranes, there are pallet-reversing operations on transfer lines and many other mobile and industrial uses. Some, in fact, may not involve motors, but instead may use either linear or rotary actuators capable of less than a full revolution—as in a ship's rudder control or the fin actuators for a ship's roll stabilization system.

The following discusses primarily wheel drives, since they must cope with virtually every demand (reversibility, overrunning load capability, etc.) found in any transmission application.

II. TRANSMISSION HYDRAULIC DIAGRAMS

A. Open-Loop System with Fixed Displacement Pump

In order to fully appreciate the advantages of a closed-loop system, the features of two simple open-loop systems will be reviewed. Figure 6-1 shows a familiar fixed displacement pump and open-center valve in an open-loop system. This

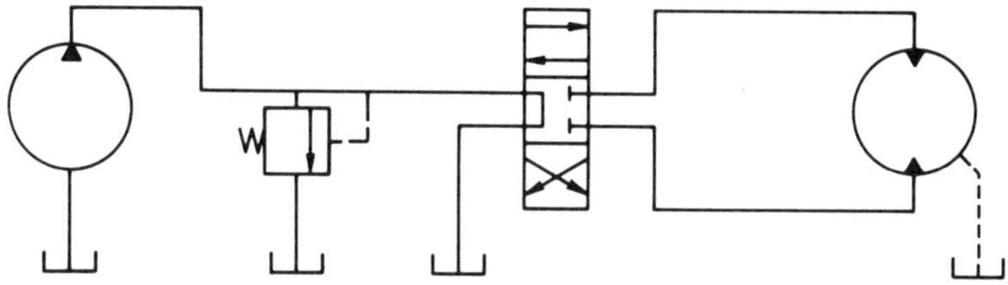

FIGURE 6-1 Open-Loop System with Fixed Displacement Pump and Open-Center Valve

arrangement is simple, inexpensive and efficient when operated at full speed in forward or reverse. Given the right kind of control valve, it is also capable of operation at partial speeds and of handling overrunning loads. The flexibility of hydraulics is, of course, retained as is the adaptability to remote control.

This system, however, has certain limitations. Power loss is high during partial-speed operation when the unused pump flow must be bypassed at full operating pressure. For a given control valve opening, the speed of the motor will vary as the load changes, since bypass flow across the valve is affected by pump operating pressure. The system also does not allow a completely smooth continuous transition in speed from forward to reverse passing through zero motor rpm.

Another shortcoming of the system is that the maximum system pressure, which limits maximum torque output of the motor, is limited to that permitted by the prime mover power capability when driving the pump at full pressure and full flow. In a fixed displacement system, there is no opportunity to trade pressure for displacement so as to stay within a constant horsepower limit.

B. Open-Loop System with Variable-Displacement Pump

The variable displacement system shown in Figure 6-2 employs a pressure-compensated (constant pressure) pump with a closed-center valve. In general,

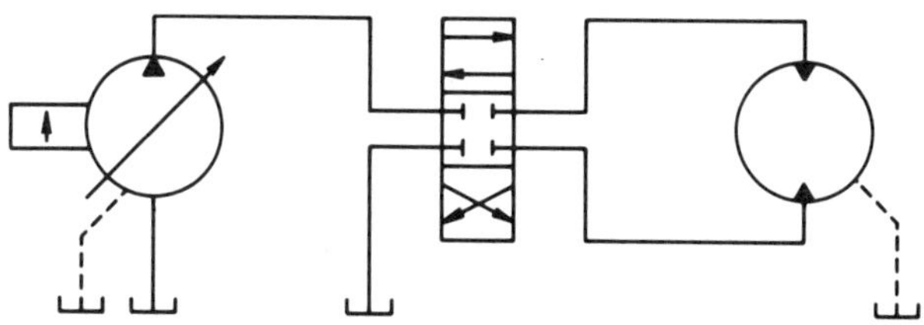

FIGURE 6-2 Open-Loop System with Variable-Displacement Pump and Closed-Center Valve

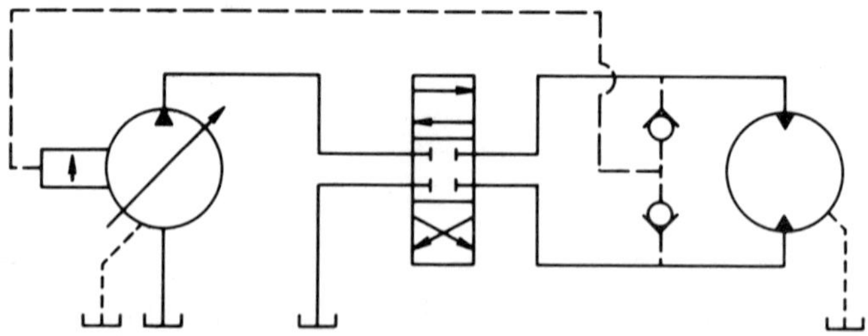

FIGURE 6-3 Open-Loop System with Load-Sensing Pump and Closed-Center Valve

given the right kind of valve, it will do everything that the fixed displacement system in Figure 6-1 will do while avoiding one of the problems. It maintains better system efficiency under partial-flow conditions, since the pump can adjust its output in accordance with the demand.

Motor rpm will still vary with load, however, since flow across the control valve will vary with pressure drop. Power loss, while less than that with a fixed pumping system, will also remain relatively high when the load is low.

These two deficiencies can be overcome by using a load-sensing pump control as in Figure 6-3. The load-sensing pump control will maintain constant flow across the system control valve for any valve opening, regardless of load pressure. It also conserves power by maintaining a pump pressure only slightly above the system needs at any time. The fact that it will be at a low, unloaded pressure during standby operation is another plus compared to the plain pressure-compensated pump, which will remain at full pressure while at zero flow.

Neither of these variable pump systems in Figure 6-2 or 6-3 will offer a completely smooth transition in speed through zero rpm, however. Nor will either of them permit a tradeoff between pressure and flow, along a constant horsepower curve, unless a torque limiting pump control is used.

C. Closed-Loop Hydrostatic Drive

The full benefits of a hydrostatic transmission are best obtained with a closed-loop system of the type shown in Figure 6-4. A variable pump, capable of pumping in either direction, is directly ported to a bidirectional motor. This closed loop is supercharged by a small auxiliary pump which is usually mounted directly on the larger variable unit. A combination of check valves connects the auxiliary pump output to the low-pressure side of the closed loop. The supercharge pressure level is established by a supercharge relief valve.

The presence of the auxiliary supercharge pump adds a small, steady parasitic loss to the system. The size of the auxiliary pump is small, however, com-

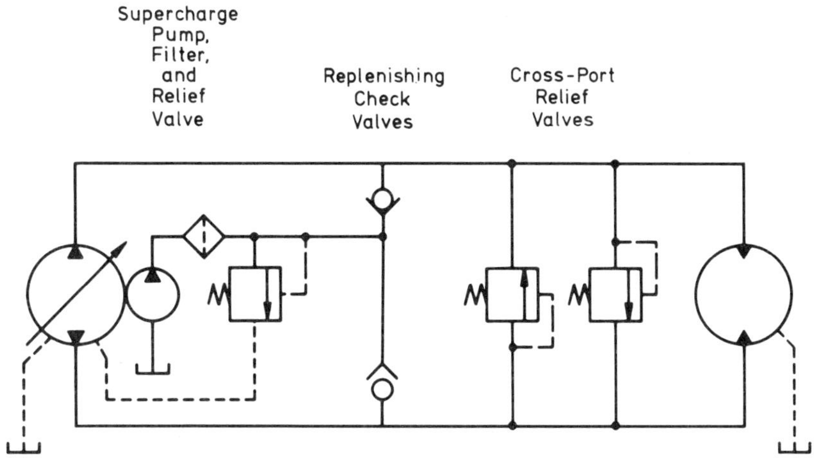

FIGURE 6-4 Closed-Loop Hydrostatic Drive

pared to the main transmission pump, since it need only provide enough flow to take care of pump and motor internal leakage which is normally low in the types of units used in transmissions. The supercharge pressure level is also low, usually about 100 psi (7 bar), so that the parasitic power loss is well under 1% of the full power capability of the transmission.

1. Benefits of Supercharge System In addition to ensuring positive transmission response when pump displacement is changed, the supercharge system has other beneficial characteristics. The presence of a positive pressure at the transmission pump inlet prevents cavitation during high-flow operation and permits the pump to be run at a higher rpm than would be possible with atmospheric inlet. For a given power level, a physically smaller pump can therefore be used.

A second benefit of even greater importance is that the presence of a supercharge pump permits the transmission loop to operate on filtered fluid by adding a small filter at the auxiliary supercharge pump outlet. This can have a positive effect on the life and reliability of the transmission.

The supercharge pump is normally also used to provide cooling flow for the main transmission pump by porting excess supercharge flow through the main pump case. This is shown schematically in Figure 6-4. This auxiliary cooling may be needed to prevent overheating of the transmission pump during extended periods of zero flow/zero pressure standby operation when there is no normal internal pump leakage.

A final benefit that comes with the presence of the supercharge pump is that it becomes easy to provide an auxiliary source of power for some secondary machine functions. This is often done by using an auxiliary pump having a flow capability beyond that required for the supercharge function alone and making the excess flow available for other uses. A priority valve arrangement ensures that the transmission loop is not starved.

2. Characteristics of Closed-Loop System Operation The closed-loop system allows smooth transition from forward to reverse through zero rpm and also allows full hydrostatic braking action in either direction. To prevent excessive loop pressure if there are high inertia loads on the motor and pump flow is suddenly reduced, crossport relief valves are commonly used as shown in Figure 6-4. As a safeguard, the same type of pressure limiting arrangement would normally also be employed in the systems shown in Figures 6-1, 6-2 and 6-3.

In addition to its ability to provide a smooth, stepless transition in speed, the closed-loop system will provide a nearly constant motor speed for any pump displacement regardless of load. This is because the types of units used–usually piston units–almost invariably have high volumetric efficiencies. The amount of internal leakage which controls the "hydraulic slip" is low at any operating pressure.

3. Torque-Multiplying Capabilities of Closed-Loop Systems Another characteristic of the closed-loop transmission is that it lends itself readily to torque multiplication that occurs when the transmission pump is run at low flow but high pressure, within the power capabilities of the prime mover. Figure 6-5 shows the allowable pressure and flow characteristics of a pump driven by a prime mover with a given power (i.e., torque) capability. The selection of a fixed displacement pump would be limited by the boundaries of that curve. A pump selected for high flow, (point A in Figure 6-5) would be limited to the pressure shown. The motor maximum torque which is proportional to pressure would be correspondingly low.

A fixed pump selected for high-pressure capability, (point B in Figure 6-5) would be limited in displacement and the maximum motor speed would be low. A compromise selection, such as point C would increase the speed somewhat but reduce the torque capability.

The closed-loop transmission, through the use of a variable pump, is able to operate at any point from A to B under the control of the operator. High motor torque can be developed at low speed for startup conditions or momentary high-load demands. For lower loads, high motor speeds are possible.

4. Use of Variable-Displacement Motors in Transmissions A further broadening of the range of operation is sometimes obtained by the use of a variable displacement motor together with the variable displacement pump. A simple example of the utility of such a combination is the use of two alternate dis-

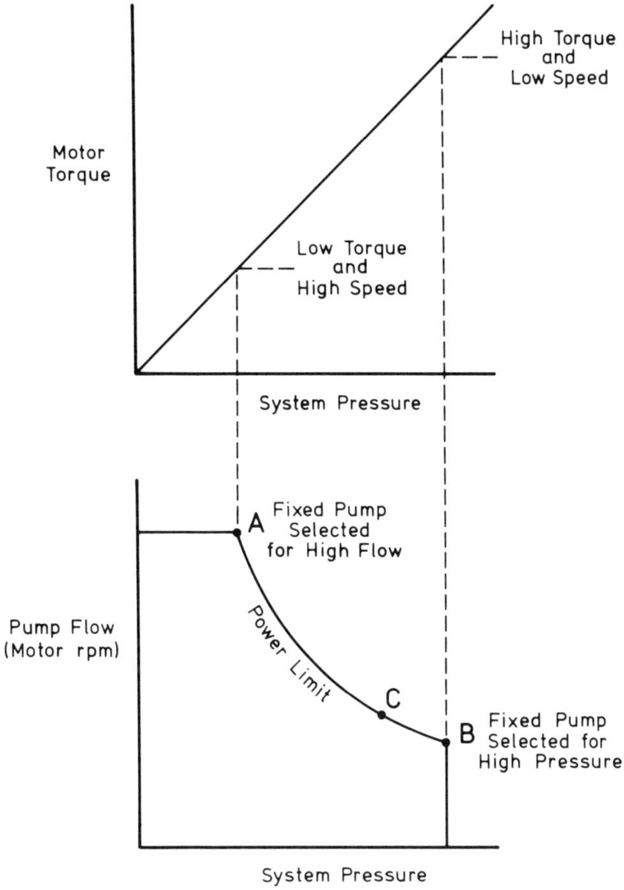

FIGURE 6-5 Closed-Loop Transmission Operating Boundaries with Variable Pump and Fixed Displacement Motor

placement positions on motors for off-the-road vehicles. The maximum motor displacement is used for maximum torque while working on rough terrain. The lower displacement (usually about 50% of maximum) is used for high speeds when traveling on smooth roads between jobs.

In some cases, the pump and motor stroke controls are synchronized so that, as the pump nears its full stroke position, the motor begins to reduce its stroke. This can give a very broad speed and torque range, as shown in Figure 6-6, and can be made to operate within the torque or power limits of the prime mover.

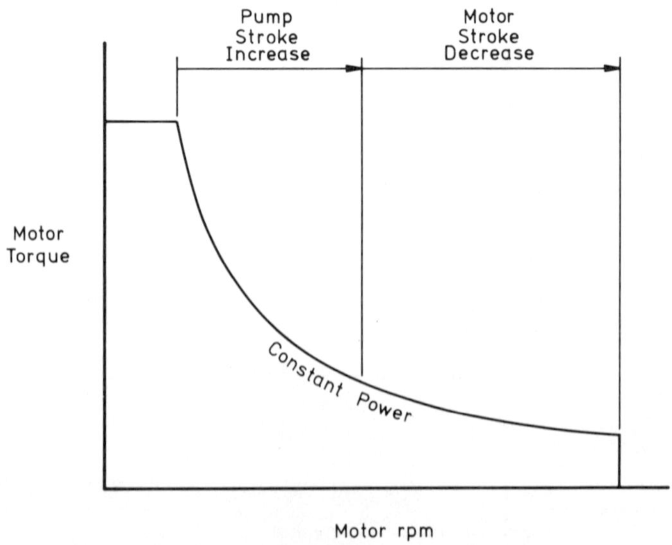

FIGURE 6-6 Closed-Loop Transmission Operating Boundaries with Variable Pump and Variable Motor

III. TRANSMISSION SYSTEM EFFICIENCY

In order to fully appreciate the high system efficiency a closed-loop transmission can attain, it is worthwhile to examine the pressure/flow diagrams of the open- and closed-loop systems. Figure 6-7 shows the open-loop, fixed-pump system of Figure 6-1. When operated at partial motor rpm and pressure as shown, the pump pressure will be only slightly higher (due to line losses) than the pressure at the motor. The unused flow will be dumped across the open-center valve, however, resulting in the losses indicated.

In Figure 6-8, the pressure-compensated variable pump system of Figure 6-2 is illustrated. Here, pump flow automatically adjusts itself to the demand, but there is a large pressure drop across the system valve in addition to line losses.

The load-sensitive variable pump system of Figure 6-3 is much better, as can be seen from Figure 6-9. The pressure drop across the system valve is limited by the load-sensing control pressure setting. There will, of course, also be some line losses.

The closed-loop system diagrammed in Figure 6-10 is clearly the best of all. There is no wasted power due to either excess flow or valve pressure drops.

These diagrams, of course, show only the losses attributable to the system and do not include those generated internally within the pump and motor. For systems of the type shown in Figures 6-7 and 6-8, however, the system losses can far exceed the component inefficiencies for many conditions of operation.

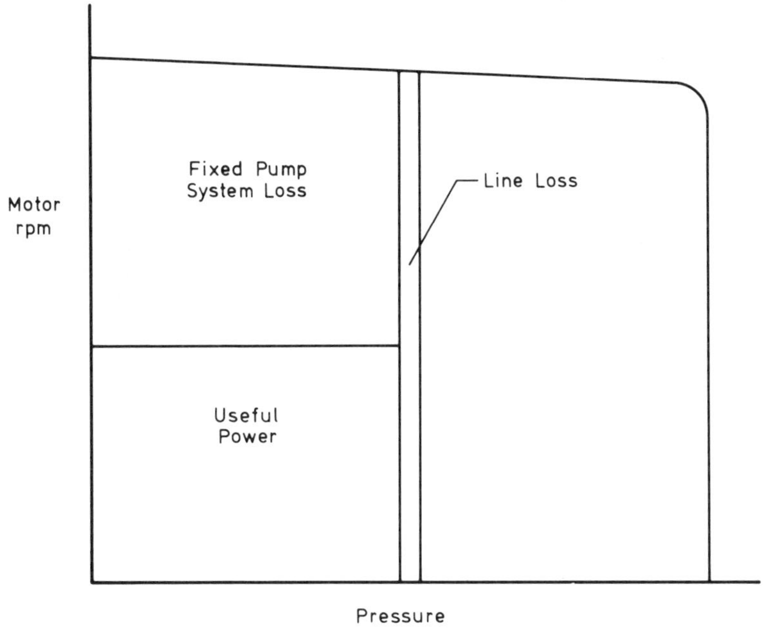

FIGURE 6-7 System Power Loss with Fixed Displacement Pump and Open-Center Valve

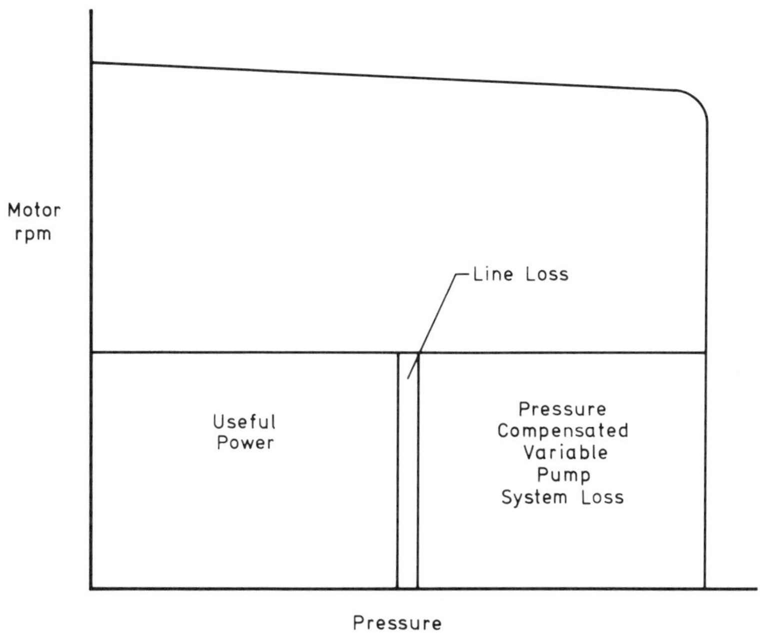

FIGURE 6-8 System Power Loss with Variable Displacement Pump and Closed-Center Valve

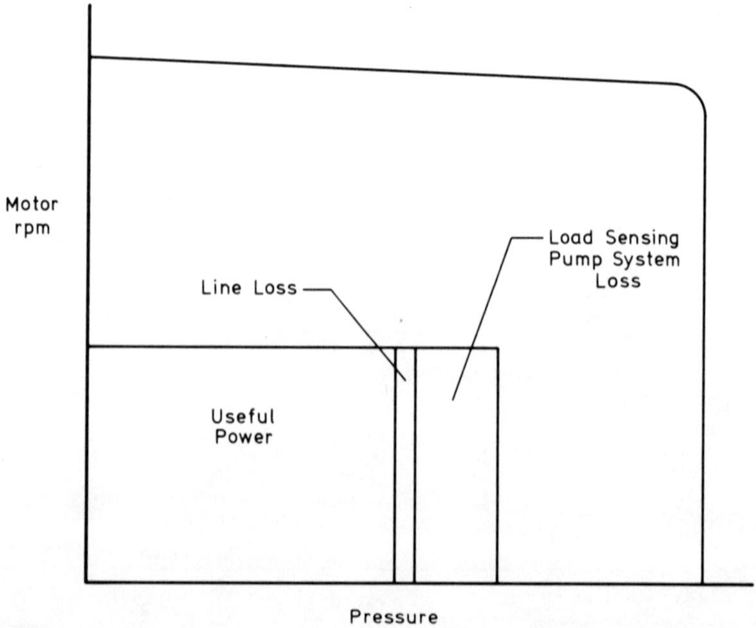

FIGURE 6-9 System Power Loss with Load-Sensing Pump and Closed-Center Valve

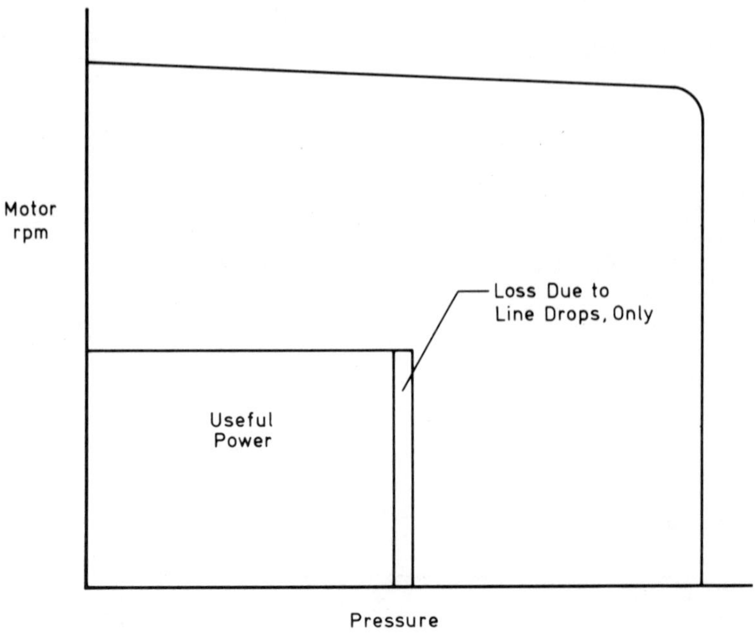

FIGURE 6-10 System Power Loss with Closed-Loop Transmission

IV. TRANSMISSION INSTALLATION CONFIGURATIONS

Most transmissions employ piston pumps of the inline type in which the direction of pumping flow is readily reversed by tilting the swash plate or yoke in the opposite direction. The compact size and high volumetric efficiency of this type of unit are also favorable for transmission use. Motors are usually of either the inline piston type, fixed or variable, or one of the high efficiency high-torque, low-speed designs.

Figure 6-11 shows a typical double-transmission pump of the type used as a wheel drive for skid-steer vehicles. There are two variable-piston pumps together with a fixed displacement supercharge pump, all driven by a common shaft. Figure 6-12 identifies the components. The two variable units are individually controllable and will drive motors at the wheels on opposite sides of the vehicle. The section between the two pumps houses the supercharge pump relief valve and check valves and also the high-pressure, crossport relief valves. The supercharge pump relief valve flow passes through the piston pump housings and out a case drainline to carry away waste heat when the pumps are operating at zero flow and pressure in a standby mode.

The pumps shown are manually controlled and capable of going across center for forward or reverse operation. When individual pump power levels

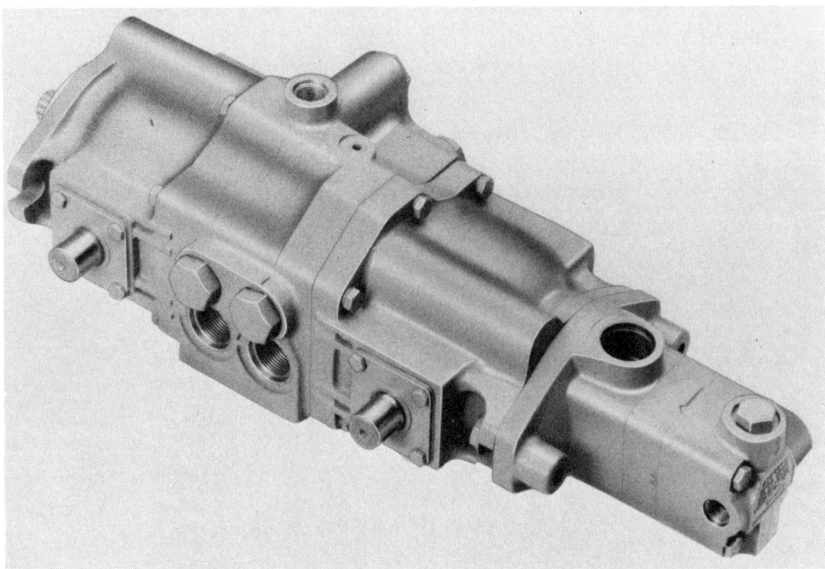

FIGURE 6-11 Hydrostatic Transmission Pump Package

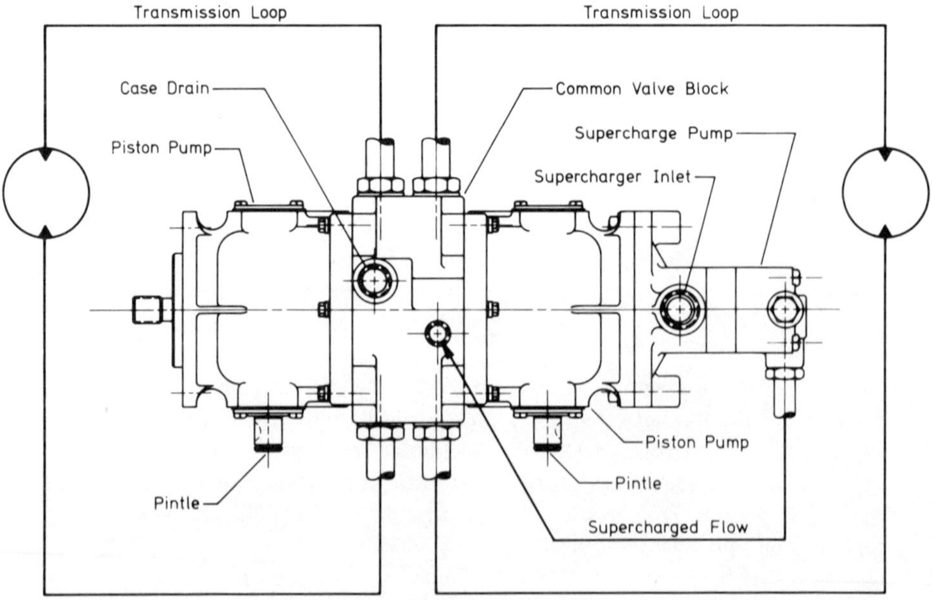

FIGURE 6-12 Hydrostatic Transmission Component Arrangement

approach the 100 hp (75 kW) range, the stroke control forces become sufficiently high so that a servocontrol is usually employed for power assist. It may also be used for installation convenience at lower power levels to facilitate remote operator control.

Figures 6-13 and 6-14 show fixed- and variable-displacement piston motors of the type typically used in transmissions. The variable motor shown has a through shaft to permit the easy installation of a disc-type parking brake. The stroke control is manual but, as in the pumps, hydraulic operation is usually employed in larger motors.

V. TRANSMISSION APPLICATIONS AND CONTROLS

In addition to wide usage in wheel drives, closed-loop transmissions see application in many other industrial and mobile functions. Operations requiring reversibility together with sensitive speed or position control are served both effectively and efficiently by this type of system. A typical system having these demands is the swing drive for rotating the boom on cranes and excavators. To meet the needs of specific applications, several types of auxiliary pump controls are available, including manual displacement control with power assist, displace-

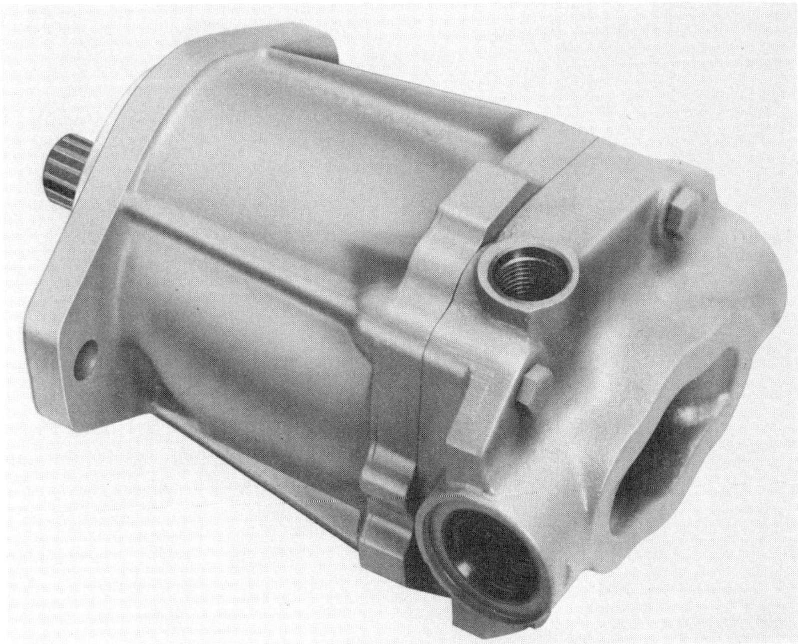

FIGURE 6-13 Fixed Displacement Piston Motor

FIGURE 6-14 Variable Displacement Piston Motor with Through Shaft

ment control with hydraulic or electrohydraulic remote control, rate and pressure limiting, torque-select control, and integrated power controls. Each one is discussed in detail.

A. Manual Displacement Control with Power Assist

A power assist system is used to reduce the operator's force level. It is directly analogous to a power steering system in an automobile. The functioning of a typical system is illustrated in Figure 6-15.

Pressure from an auxiliary pump, which could be the system supercharge pump, is directed to hydraulic cylinders which move the pump's stroking mechanism. A "follow valve" arrangement is commonly employed in which one element of the valve, in this case the sleeve, is keyed to the stroking mechanism and moves with it. The valve spool is displaced manually by the operator. In this arrangement, any displacement of the spool away from neutral results in a corresponding movement of the stroking mechanism, both in direction and in amount. Each spool position corresponds to a specific stroke position at which point the spool lands and sleeve ports are again aligned and flow to the auxiliary stroking cylinders stops. If the spool is spring-loaded toward center, release of the handle by the operator will recenter the spool and bring the transmission pump back to zero flow.

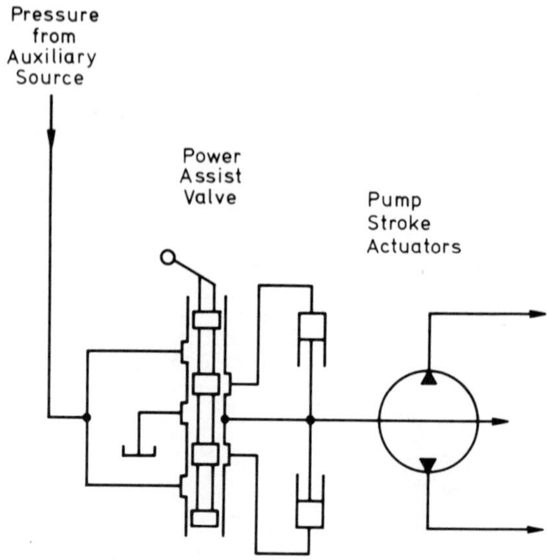

FIGURE 6-15 Manual Displacement Control with Power Assist

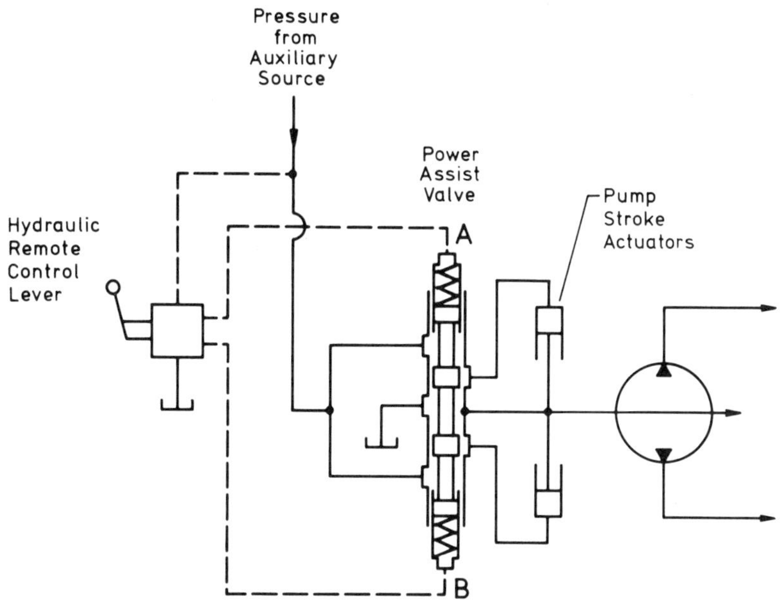

FIGURE 6-16 Displacement Control with Hydraulic Remote Control

B. Displacement Control with Hydraulic or Electrohydraulic Remote Control

To simplify remote operation, the power assist valve spool is sometimes hydraulically positioned as shown in Figure 6-16. The hydraulic remote control valve applies pressure in line A when the handle is moved forward or in line B when it is moved backward. The magnitude of the pressure is proportional to handle position. These hydraulic pressures, operating against the spring-centered valve spool, cause a displacement upward or downward that is proportional to the pressure. The stroke control mechanism then moves until the valve sleeve is recentered on the spool lands. In this arrangement, each handle position of the hydraulic remote control valve, therefore, corresponds to a specific pump stroke position.

This type of power assist control is readily adaptable to electrohydraulic servooperation. The electrohydraulic servovalve replaces the hydraulic remote control valve and is used to apply pressure at points A or B of Figure 6-16. The servovalve would normally be mounted directly on the power assist valve. For manual operation, the operator's control would be electrical, providing excitation to the servovalve corresponding in direction and magnitude to the pump displacement desired. This system, is, of course, also directly adaptable to programmed electrical or microcomputer control systems.

C. Rate and Pressure Limiting

When the transmission motor is driving a high inertia load, it is often desirable to protect the system against repeated pressure shocks due to rapid changes in pump stroke position by the operator. While a measure of protection is afforded by crossport relief valves, system operation can still be rather violent and power can be wasted or the system overheated by repeated relief valve operation.

Power assist controls are, therefore, sometimes equipped with a rate limiting feature. The rate of change of pump stroke is limited to a moderate value, regardless of the rapidity of the operator's motions. A maximum rate of increase or decrease in pump flow is established to provide the maximum rate of load acceleration (including braking) without exceeding relief valve settings.

In some systems where the load can be completely stalled (as when it is driven against a positive stop), a second feature is highly desirable. This is a pressure-limiting override control which will reduce pump stroke to zero, overriding any control signal from the operator, when the system pressure reaches a set value. It is set at a value below the crossport relief valve settings and prevents unwanted energy loss that would occur if flow were passed over the relief valve during stalled load conditions.

The pressure-limiting override has the same beneficial effect as a rate control in minimizing system pressure shocks due to rapid operator control motion. When a sudden large change in displacement is called for by the operator, the rate of stroke change is automatically scheduled by the pressure-limiting override so that the override pressure setting is maintained but not exceeded. This assures maximum acceleration or braking without exceeding the relief valve setting.

D. Torque-Select Control

The torque-select control offers a somewhat more sophisticated way to provide complete smoothness of operation together with total control by the operator. Its general functioning is illustrated in Figure 6-17.

Pressure from a hydraulic remote control or electrohydraulic servo valve is applied at points A or B against a spring-centered spool. As in Figure 6-16, each level of pressure results in a proportionate spool displacement and in movement of the pump-stroking mechanism. In this control there is no follow valve action to recenter the sleeve ports with the spool lands, however. Instead, the pressure in lines C and D of the transmission loop is applied to smaller areas at the ends of the torque-select control valve. Pump stroke continues to increase in the selected direction, until the magnitude of the pressure differential between C and D overcomes the hydraulic remote control pressure and recenters the torque-select control valve.

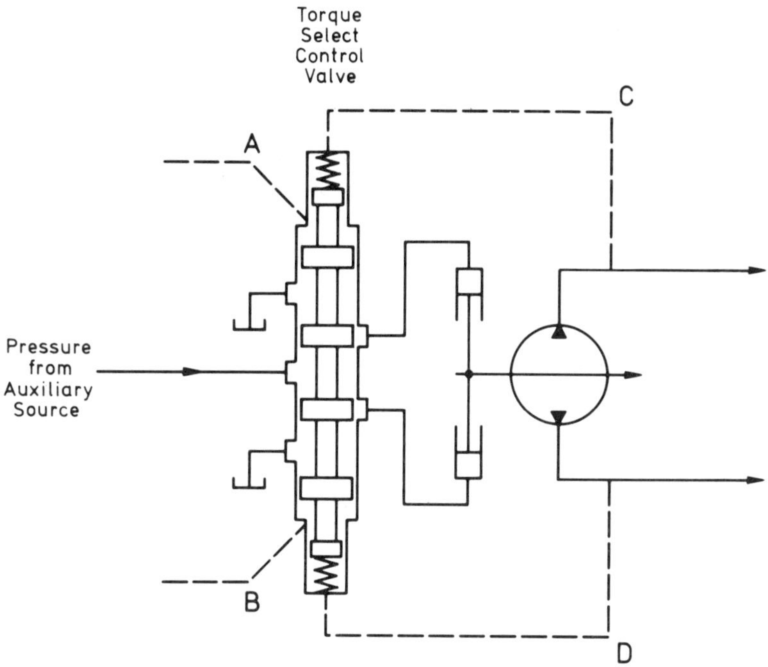

FIGURE 6-17 Torque-Select Control

In this type of control, each value of hydraulic remote control pressure produces a corresponding value of pressure differential at the transmission motor. Since motor output torque is proportional to the pressure differential, each position of the hydraulic remote control handle corresponds to a specific value of torque delivered by the motor to the load.

In systems such as swing drives where the load is primarily inertial, each remote control handle position produces a given rate of acceleration of the load. In the control shown in Figure 6-17, recentering the control handle once the load is in motion allows the load to coast or freewheel at constant velocity, diminishing only due to friction effects. To bring the load to rest, the direction of the handle is reversed, applying a braking torque proportional to handle position.

The operation of this control provides complete smoothness of acceleration or braking at a rate that is under the full control of the operator. If the freewheeling center position is for some purpose not desired, an option of neutral braking at some preselected rate is available.

The torque-select control has some interesting features. It can be used, for example, to provide nearly constant tension in a winch line, automatically reeling line in or paying it out without a change in the operator's control setting. In an excavator swing drive, it allows the operator to exert a constant, selected value of force of the bucket against the wall of an excavation. In the swing drive of a crane, when the remote control handle is in the neutral position, the boom will automatically center itself over the load when cable tension is applied.

Like the hydraulically controlled displacement control, the torque-select control is directly adaptable to electrohydraulic servooperation and microcomputer control.

E. Integrated Power Controls

Many vehicles that use hydrostatic drives operate with a constant governed engine speed. The speed is set at a value that will permit the generation of full engine power when the work cycle demands it. However, in normal working modes, the power demand may be less than half the full-power capability of the engine.

When this is true, the operation of the engine at high-rpm and low-throttle setting (low-engine torque) results in less than optimum fuel consumption. A saving of from 5% to perhaps as much as 20% is theoretically possible by reducing rpm as the power demand decreases. Fuel consumption is improved by operating at higher engine torque and lower speed.

The development of microcomputers and of sensors to monitor engine and transmission conditions makes integrated power management possible. The simplicity of application of electrohydraulics to transmission displacement controls is the other necessary link in the chain.

Index

P